E. Gebhart · R. M. Arutyunyan

Anticlastogens in Mammalian and Human Cells

With 16 Figures and 19 Tables

Springer-Verlag
Berlin Heidelberg New York
London Paris Tokyo
Hong Kong Barcelona
Budapest

Professor Dr. ERICH GEBHART
Institut für Humangenetik
der Universität Erlangen-Nürnberg
Schwabachanlage 10
8520 Erlangen, FRG

Professor Dr. RUBEN M. ARUTYUNYAN
Chair of Genetics and Cytology
Department of Biology
State University of Yerevan
Yerevan 374049, USSR

ISBN 3-540-53357-5 Springer-Verlag Berlin Heidelberg New York
ISBN 0-387-53357-5 Springer-Verlag New York Berlin Heidelberg

Typesetting: International Typesetters Inc., Makati, Philippines
31/3145-543210 – Printed on acid-free paper

Preface

Antimutagenesis, because of its intricate relationship with anticarcinogenesis and its significance in genetic disease, but also in cancer and perhaps even in aging, has experienced a renaissance during the last 10 years. Its roots, however, go back to the fifties, when Novick and Szilard created the term "antimutagenesis", and when protection by antimutagens also of chromosome structure from the detrimental effects of ionizing radiation and certain chemical mutagens was reported first.

Indeed, an important part of all efforts in the field of antimutagenesis research was contributed by studies on anticlastogenesis, i.e. the reduction of the amount of chromosome damage induced by mutagens. Both authors were involved in this type of research performed on human and mammalian cells from its very beginning.

As a lucky chance (i.e. a visiting grant of Deutscher Akademischer Austauschdienst – DAAD) led Ruben Arutyunyan to my lab, we decided to invest our years of experience into this summarizing book. The result of this cooperation, naturally cannot be absolutely exhaustive, as we had to try a short summary of all data available to us, but also to concentrate on selected aspects of anticlastogenesis research. Obviously such a selection will turn out to be subjective, but we seriously exerted ourselves to cover as many aspects of anticlastogenesis as possible. We hope that our original intention will be fulfilled by our book, i.e. to draw the attention of many scientists and students to this advanced field of research which, nevertheless, still seems very promising to us, and to advance understanding of the complex processes and interrelationships of anticlastogenesis, and antimutagenesis.

It is our desire to give our thanks to all our colleagues who allowed us to include into this book their unpublished data or parts of their publications, but also to those who encouraged us by valuable advice and discussions. We greatly acknowledge the financial support of R.A.'s stay in Erlangen by DAAD which was a basic requirement for writing this book. Last, but not least, we are also indebted to the publisher for the extremely good cooperation.

Erlangen, May 1991

Prof. Dr. E. GEBHART
(on behalf of the authors)

Contents

Abbreviations

AET	β-Aminoethylisothiouronium
AMSA	amsacrine
Ara-C	cytosine arabinoside
BaP	benzo$\langle\alpha\rangle$pyrene
BLM	bleomycin
BrdUrd	5-bromodeoxyuridine
CHO	Chinese hamster ovary cell line
COL	colchicin
DEB	diepoxybutane
DMBA	dimethylbenz$\langle\alpha\rangle$anthracene
DNA	deoxyribonucleic acid
EMS	ethylmethane sulfonate
G0	stationary stage of interphase
G1	prereplicative phase of interphase
S	replication phase of interphase
G2	postreplicative phase of interphase
h	hour
HCT	homocysteine thiolactone
HN2	nitrogen mustard
8-HQS	8-hydroxyquinoline sulfate
ICPEMC	International Commission for Protection Against Environmental Mutagens and Carcinogens
M	Molar
MMC	mitomycin C
MMS	methylmethane sulfonate
MNNG	N-methyl-N′-nitro-N-nitrosoguanidine
NAC	N-acetylcysteine
NaF	sodium fluoride
NHAAF	N-hydroxyacetylaminofluorene
NMU	nitrosomethylurea
NQO	4-nitroquinoline-1-oxide
PCC	premature chromosome condensation
PCZ	Procarbazin
PVP	polyvinylpyrrolidone
SCE	sister chromatid exchange
TEM	triethylenemelamine

TEPA	triethylenephosphamide
ThioTEPA	triethylenethiophosphamide
TPA	tetradecanoylphorbolacetate
TR	trenimon
UV	ultraviolet
VINC	vincristin
WR 2721	S-2-(3-aminopropylamine)ethylthiophosphate

1 Introduction

There is no longer any doubt about the ability of ionizing radiation, or of a series of chemicals to induce changes in genetic material (mutations). The practical meaning of induced mutations is substantiated by the manifold evidence of their detrimental consequences in man. Mutations in germ cells can cause a large variety of genetic diseases which have been (more or less) well defined by medical genetics (McKusick 1988). Far more than 4000 monogenic hereditary traits have been listed in this edition of the catalogue of *Mendelian Inheritance in Man*, most of them connected with more or less serious illness. As shown in Table 1, the "burden" of diseases caused by changes in genetic material is quite considerable. It would be much more impressive if all the diseases were added, which mutations inflict on humans in a multifactorial way, i.e. by interaction of changes in the genetic material with environmental factors. Examples for the latter are diabetes, hypertonia, several heart diseases, various kidney diseases, allergies and related diseases, and many others (see literature on human genetics: e.g., Lenz 1979; Thompson and Thompson 1986; Vogel and Motulsky 1986; and many others).

Moreover, mutations in the genetic material of somatic cells can also lead to dramatic consequences. It has been well established now that the complex, multistep process of malignant transformation and cancer development is causally related to various types of mutations. This same process can also be experimentally induced (Miller 1978; Farber 1984; Weinstein et al. 1984; Barbacid 1986; Upton 1986; Slaga 1989; Fearon and Vogelstein 1990; Gebhart 1990a). Very extensive and effective research in the fields of classical cytogenetics and molecular genetics of cancer cells over the past two or three decades has produced a rather concrete picture of the events and mechanisms rendering a normal cell malignant.

Mutational changes of so-called cellular oncogenes apparently play a fundamental role in cell transformation and tumor progression. These oncogenes, by their nature, are highly conservative, and, therefore, important genes normally coding for a series of specific proteins particularly involved in growth control and differentiation processes. More recently, the basic importance of control genes ("tumor-suppressor" genes) was detected. Mutational changes in these genes, causing a loss of their control functions, can also lead to malignant transformation. Table 2 presents the main mutational mechanisms considered to be involved in malignant processes at our present state of knowledge. Here, the basic role that chromosomal changes play in the process of malignancy should be mentioned (Sandberg 1980, 1990; Mitelman 1985, 1988; Heim and Mitelman 1987; Gebhart 1990b).

Table 1. The human genetic "burden"[a] (After Brusick 1980)

Mutational change	Affected individuals per 1000
Monogenic traits	
Autosomal dominant gene mutations	1.85–2.64
Autosomal recessive gene mutations	2.23–2.54
Sex-linked gene mutations	0.78–1.99
Chromosomal anomalies	
Numerical aberrations	3.0
Structural aberrations	2.1
Variants	1.7
Total load	11.7–14.0

[a]An additional 10–30% of congenital malformations and other complex disorders are believed to be the result of genetic defects. Among 3714 spontaneous abortions 40.4% were found to be affected by a chromosomal anomaly, in most cases a numerical one (cited from Sperling 1984).

Table 2. Mutational mechanisms involved in the process of malignant transformation

Type of mutation	Main target genes	Consequences
Point mutations	Proto-oncogenes	Undue expression (activation)
	Suppressor genes	Loss of heterozygosity, inactivation
Chromosomal rearrangements	Proto-oncogenes	Activation, overexpression
	Control genes	Loss of function, undifferentiated growth
Chromosome deletion/ monosomy	Suppressor genes	Hemizygotization of mutations in suppressor genes
Gene amplification	Proto-oncogenes	Uncontrolled expression (mainly overexpression)
	Specific genes of metabolism	Generation of therapy resistance

In addition, other somatic processes like aging, teratogenesis, and the multifactorial pathogenesis of other diseases seem to be related to mutational changes in somatic cells. All these findings and evidence necessitated a serious search for possibilities to protect the human genome from these adverse influences.

It has been recommended for many years to prohibit mutagenic agents from the human environment by warning against their use and against their distribution (e.g. Barthelmeß 1959; Hollaender 1971; Ehling 1976; Sorsa and Vainio 1982). However, since a general and absolute prohibition of mutagenic agents is not realistic in many fields (e.g. natural contamination of our food by mutagens; specific working places), or would even be detrimental in some cases (e.g. life-saving cancer therapies or immunosuppressive therapies), additional possibilities of protecting the genetic material of exposed individuals from deleterious effects due to environmental mutagens had to be examined.

The first steps toward antimutagenic protection were initiated in the field of radiation biology when substances were detected which efficiently protected biological systems from the toxic effects of ionizing radiation. The same or similar substances, later on, also proved to be active in reducing the mutagenic effects of irradiation (e.g. Mikaelsen 1952, 1954; Riley 1952). These substances, therefore, were called antimutagens, a term which, to the best of our knowledge, was first used by Novick and Szilard (1952) to describe the reduction of spontaneous or chemically induced mutagenic damage in bacteria.

Since the action on the genetic material is very similar between ionizing radiations and certain chemicals, it was consequently presumed that antimutagens should also reduce the damaging activity on the genetic material of those chemicals. This assumption was experimentally verified already in the 1950s in bacteria (Novick and Szilard 1952) and in cytogenetic test systems (Avanzi 1957; Moutschen 1958, 1960). Despite the technical problems hampering the progress in antimutagen studies, extensive reviews on antimutagens, were published in the 1970s (e.g. Gebhart 1974; Clarke and Shankel 1975). Antimutagen research experienced a renaissance in the 1980s when the interest of researchers increasingly focused on natural mutagens. A first International Conference (Shankel et al. 1986) summarized the results obtained from noncytogenetic systems. With the growing evidence of an extremely close relation between mutagenesis and carcinogenesis, the aspect of protection was extended to anticarcinogenesis (reviewed by De Flora 1988).

From the beginning, cytogenetic test systems have been involved in antimutagen research, a fact which has been neglected by several of the more recent reviews and conferences. However, the protection of chromosomes from the damaging activity of known mutagens has been under study since the 1950s using the well-defined chromosomal aberrations. Later, sister chromatid exchanges and micronuclei were also used as endpoints of evaluation. Considering the immense practical impact of chromosomal anomalies on human health in terms of human chromosomal diseases and, in particular, in terms of tumor cytogenetics, the data obtained from these studies gained direct practical importance. Substances reducing the amount of chromosomal damage induced by mutagens were now termed "anticlastogens" (Gebhart 1972, 1974).

The complexity of the obtained data, as pointed out by Waters et al. (1990), "is due in part to three simultaneous experimental variables: the mutagens (and/or spontaneous mutation) studied, the antimutagens studied, and the short-term tests used."

To better understand the problem of anticlastogenic action, therefore, it seems to be necessary to first analyze the developments and trends in the studies on the clastogenic action of chemicals in the last years.

A definite part of the great number of chemicals composing our environment was classified on the basis of their mutagenic activity. Genetic screening and monitoring programs became an important part of the obligatory control of newly synthesized drugs and other chemicals. In the 1970s a new interdisciplinary science was established, "genetic toxicology", a designation which indicates its purpose. Scientists agreed that though the research on chemical mutagens is even more complicated than that on radiation mutagenesis, it is indispensible for practical reasons and yields manifold and extraordinary information on the mechanisms of mutagenesis. The practical importance of chemicals regarding environmental exposure made it necessary to focus research interests on chemical mutagenesis.

In addition, based on simple in vitro test systems with the addition of chemicals directly to (mammalian or human) cell cultures, the subsequent registration of the induced cytogenetic effects, and on acceptable hypotheses of the underlying molecular mechanisms (e.g., Kihlman 1966; Rieger and Michaelis 1967; Bender et al. 1974), tremendous progress was made toward new model ideas on the formation of chromosomal changes. Furthermore, the mechanisms of modulation of the induction of these cytogenetic changes by inhibitors and inductors of metabolic pathways and of DNA repair were successfully studied in these systems. These findings and evidence, however, would not have been obtained if there had not been a better understanding of the metabolic pathways acting within the target cells, tissues, and organisms (Ames et al. 1973; Vogel et al. 1974; Glatt and Oesch 1986), and thus interacting with the applied anticlastogens. Together, these facts allowed a deeper molecular understanding of the induction and modification of cytogenetic changes in cells of higher organisms. Beyond the classical chromosome aberrations, experimental cytogenetics introduced some further important approaches to modern mutation research, e.g., the analysis of sister chromatid exchange (SCE: Latt 1973, 1974; Perry and Wolff 1974), the estimation of the number of induced micronuclei (Matter and Schmid 1971), and also the possibility to study mutagenic action directly in the interphase by using the technique of premature chromosome condensation (PCC: Hittelman and Rao 1974). It is worth mentioning that some of these "endpoints" were detected or developed many years before they became a practical tool of experimental cytogenetics in the 1970s.

Besides these new methods and molecular approaches, a more quantitative approach to the understanding and interpretation of the mechanisms of chemical clastogenesis was created which previously had been well developed in radiation mutagenesis. Thus, mathematical modeling became one of the most reliable methods in this regard.

Problems pertaining to the identification and control of environmental mutagenesis are also confronted with the reality of the combined influences of various agents on the studied target cells or organisms. Moreover, the direct action of a "pure" mutagen is rather unique in a natural environment, but, in most cases, has no practical significance. Any action of the mutagen in nature is modified either by factors acting on it in the sense of potentiation or inhibition, or detoxification. There is a large series of intra- and extracellular inhibitors at various levels of mutagenic action (De Flora and Ramel 1988).

For the detection and definition of anticlastogens correct qualitative and quantitative estimations of their effects are necessary. One of the main problems in the qualitative approach is the need for adequate test systems in all their complexity to solve concrete tasks. With such systems the negative or positive effects, of a particular, presumed anticlastogen can be estimated. This qualitative research can be made in a comparative way by using test systems or by considering cytogenetic aberrations. However, a quantitative prediction of the influence on the experimental conditions changed by the studied clastogen and/or anticlastogen is not taken into account. Therefore, comparisons of the qualitative outcome of experiments on different cell cultures or organisms have to be made very cautiously.

Another problem, which also complicates extrapolation of experimental data, is the comparison of in vivo and in vitro systems, unless defined quantitative principles of extrapolation are used. Therefore, in the following chapters diagrams will be presented that can be particularly useful in any of these systems.

Quantitative estimates of the effects of anticlastogens can be obtained only when correct statistical design, analysis, and interpretation of data are available, as proposed by Whorton et al. (1984). A quantitative approach to the estimation of anticlastogenic efficacy will be presented that has shown its viability and reliability in in vitro and in in vivo studies (e.g. Bochkov and Chebotarev 1989). Moreover, diagrams for decision-making on quantitative grounds in the screening of anticlastogens will be presented.

The need to summarize information on anticlastogenic action, which could be extrapolated to human risk populations, directs our concentration on test systems involving mammalian and human cell material and on the separation of in vitro from in vivo observations. The experimental data obtained from these systems yield the most direct possibility of extrapolation. In addition, the practicability of such short-term systems was one of the main reasons for the fact that most data on anticlastogenic action have been obtained most recently from tests on cell cultures. Since genetic material is universal and the main principles of antimutagenesis are valid for all organisms, it seems reasonable to focus on mammalian and human cells. There are, however, some very specific difficulties regarding extrapolation which could cause false-positive or false-negative results. They are mainly due to metabolic differences in various systems as compared to the human organism. Therefore, one can base general predictions on the mutagenic or antimutagenic action of certain substances in short-term tests, but "... one should remember that the surface of an agar plate cannot be used as a model for a whole animal; particularly, no account can be taken of pharmacokinetic and pharmacodynamic parameters" (Garner 1980).

Following the first detection of radiation antimutagenesis, the problems involved in the extrapolation of experimental findings on man and in its practical application were recognized. There are several levels of extrapolation involved, for instance from in vitro to in vivo conditions, from somatic to germ cells, from chromosomal to general genetic damage, or from animal models to man. Several models have been proposed thus far to render these extrapolation steps as realistic and reliable as possible (e.g. Bridges 1980; Sobels 1980). In addition, specific ways of extrapolation have been examined concerning the comparison of sister chromatid exchange and gene mutations (Carrano and Thompson 1982) in relation to the quantitative approach (Bochkov and Chebotarev 1989). Though the principles of this extrapolation have now been established, there are no comparable examples in the field of anticlastogenesis. In regard to the problems of anticlastogenic action, we are also confronted with the question as to how realistic the practical use of antimutagens is in man. Considering the extensive findings on natural antimutagens (e.g., Ames 1983, 1984), which are often part of our daily food (Namiki and Osawa 1986), this question is easily answered. Natural antimutagens have been an important part of the nutrition of mankind for thousands of years and are significant for the mutation process in microevolution. Therefore, to recall these natural tools of antimutagens can even stimulate their normal uptake. The fear of inhibiting the mutation process in man, i.e., an end to microevolution, seems to be unreasonable to us. Besides the normal uptake of antimutagens, the use of high doses of natural or synthetic antimutagens could be considered acceptable for individuals with a high risk of exposure to mutagens, with respect to individual prognosis, as will be shown below.

On the basis of modern understanding of the mechanisms of clastogenic and anticlastogenic action, the present state of knowledge in this important field of research will be presented in the following. In addition, the interpretation of the results from various test systems and the various endpoints of cytogenetic action will be discussed comparatively in great detail as well as the technical problems and shortcomings of these studies. Furthermore, several practical consequences regarding further experimental research and practical application in high risk individuals will be presented. Finally, the perspectives of anticlastogen research will be considered with respect to research and application.

2 Principles of Clastogenic Action and Its Estimation

A major prerequisite for understanding the mechanisms of anticlastogenesis is knowledge of the nature and mechanisms of induction of chromosome damage. Until now no unified theory exists on the formation of chromosome aberrations in cells of higher organisms, although several hypotheses have been more or less well documented.

Some of them date back as far as 1932 when Stadler first tried to explain the nature of induced mutations in plants or to 1940 when Sax first developed the so-called break-reunion hypothesis. According to Sax, all structural chromosomal aberrations are derived from spontaneous or induced chromosome or chromatid breaks. This hypothesis, in principle, is still valid today and has recently experienced very elegant support by studies on the action of specific restriction enzymes on chromosomes; this will be discussed below (Obe et al. 1987). An alternative to the break-reunion hypothesis is the so-called "exchange hypothesis" (Revell 1955), which denotes that all detectable aberrations are derived from primary exchange processes within or between the chromosomes. However, modern molecular theories on the formation of chromosomal aberrations on the basis of DNA single and double-strand breaks (see Bender et al. 1974; Natarajan 1984; Nowak and Obe 1984) favor the "break-reunion hypothesis."

2.1 Molecular Mechanisms of Clastogenic Action

A general model for the production of aberrations in eukaryotic chromosomes was proposed by Bender et al. in 1974, the basic assumptions of which were (1) the chromosome is mononeme, containing only a single DNA double helix before replication; (2) aberrations result only from lesions in DNA; (3) the primary lesions may be either polynucleotide strand breaks or altered bases which are converted to strand breaks either by enzymatic repair mechanisms or as a consequence of normal, semiconservative DNA replication; (4) the final expression of polynucleotide strand breaks as chromosomal aberrations of various types is the consequence of repair mechanisms, DNA replication, and single-strand DNAse. This model was shown to be consistent with the then available data on the production of chromosomal aberrations induced by a variety of chemical agents as well as by ionizing radiation. In particular, it accounts for the observed differences in the types of aberrations produced by these two factors at different stages of the cell cycle.

As DNA is the main target of any mutagenic change, it is now generally accepted that clastogens display their damaging activity also via their reaction with DNA. Thus far, many molecular pathways have been shown to be involved in these interactions and, in addition, complex repair systems of damaged DNA are superimposed on the primary changes. The latter "... include DNA strand breaks (ionizing radiations, UV, bleomycin), pyrimidine dimers (shortwave UV), base alkylation (alkylating agents), inter- and intrastrand cross-links (polyfunctional alkylating agents, psoralene, and longwave UV), and intercalation (acridine, proflavin, etc.)" (Natarajan 1984; Fig. 1).

Alkylating agents, for instance, have been used as model substances in comparative studies on the reactivity to nucleophiles (Vogel and Natarajan 1982). It was shown that at equivalent toxic levels, agents with high S-values (S = the

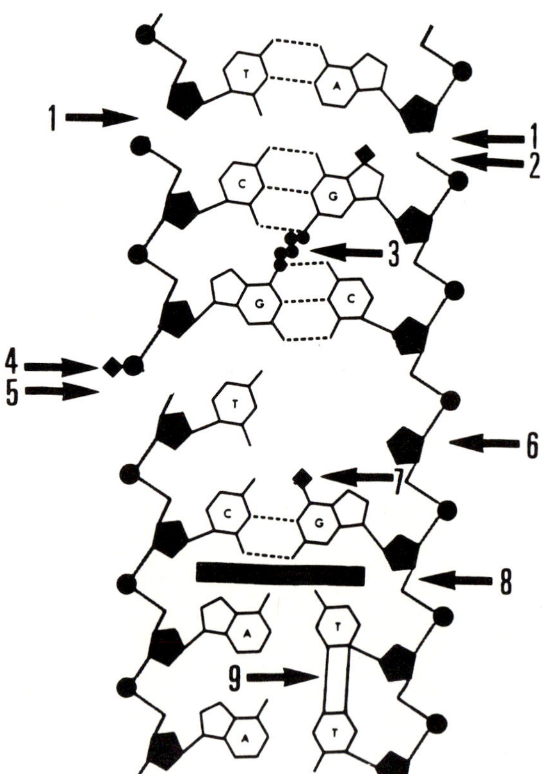

Fig. 1. Major targets in DNA of the action of clastogens. *1* Double-strand break; *2* radical reaction; *3* interstrand cross-linkage; *4* formation of phosphotriester; *5* single-strand break; *6* apurin ic site; *7* base alkylation with loss of hydrogen bond, *8* intercalation; *9* thymidine dimerization

Swain-Scott substrate constant) induced more chromosomal aberrations than those with low S-values. "Since alkylating agents with low S-values alkylate 06 of guanine more efficiently in comparison to alkylating agents with high "S", it was concluded that ... other lesions such as N7 guanine may lead to chromosome aberrations and SCEs" (Vogel and Natarajan 1982). These authors also indicated other important parameters pertaining to the mutagenicity of alkylating agents: "Examples are kinetics of decomposition, uptake and distribution of the substances within the organism, reactivity under different experimental conditions, and the effect of methylation vs. ethylation." Later, Natarajan et al. (1984) confirmed the dependence of the cytotoxic (clastogenic) activity on five monofunctional alkylating agents from their S-values.

Valuable contributions to the understanding of the molecular mechanisms involved in the formation of chromosomal aberrations were made by the groups of Obe and Natarajan using restriction endonucleases. They could show that the double-strand break of DNA is the ultimate lesion for the induction of chromosome damage (Nowak and Obe 1984; Natarajan and Obe 1984; Obe et al. 1987). As mentioned above, these data clearly favored the break-reunion hypothesis of Sax as compared with Revell's exchange hypothesis. It should be noted that restriction endonucleases, which induce blunt ends in the DNA strands, did not induce more chromosomal aberrations than endonucleases cutting via cohesive ends (Obe et al. 1987). The differential staining of chromatids using the BrdUrd-labeling technique (see Sect. 2.3 for details) in cells treated with restriction endonucleases revealed that the M2 metaphases did not exhibit increased numbers of chromosome aberrations after treatment. Only the first posttreatment metaphases had increased aberration frequencies (Gustavino et al. 1986). The potential of *Neurospora* endonuclease to transform DNA single-strand breaks into double-strand breaks became a convenient technique to analyze the mechanisms of action of various mutagens.

By using this technique, it is possible to distinguish agents, which induce lesions that can be converted into double-strand breaks, from those which originally cause double-strand breaks (Natarajan and Obe 1984). The restriction endonucleases themselves act directly (S-phase independent) as chromosome breaking agents (Obe et al. 1987). However, the data obtained from studies on the action of restriction endonucleases emphasize to what extent the combined effect of mutagens depends on their molecular mechanisms of action. In Chinese hamster ovary (CHO) cells posttreatment with *Neurospora* endonuclease of bleomycin (S-phase independent)-treated G_2 cells increased the levels of chromatid-type aberrations. But the same posttreatment produced no effect on the cells treated in G_2 with an S-dependent mutagen that does not directly increase the frequency of single-strand breaks, e.g. mitomycin. For monofunctional alkylating agents, like methyl methanesulfonate, which induce single-strand breaks, posttreatment with endonuclease under the same experimental conditions induced chromatid-type aberrations (Nowak and Obe 1984). Obe et al. (1987) noted that "these enzymes act like scissors which can be used to cut the chromosomes and thereby allow us to gain more insight into the mechanisms by which chromosomal aberrations are formed."

The role of DNA repair is extremely important for all mechanisms of exogenous action on the cell, including cytotoxic effects, and the induction of chromosomal aberrations and SCE. The DNA repair systems important for clastogenesis involve photoreactivation, excision repair and postreplication repair (for a more detailed description of the role of each system, see Sasaki 1982).

Hanawalt (1989) enumerated some historical landmarks, i.e., the year of detection, in the understanding of DNA repair in a special table. For eukaryotes he specified: photoreactivation (1960); pyrimidine dimer excision, repair replication (1967); UV-sensitive mutants (1968); "Weigle" reactivation/mutagenesis (1970); postreplication repair; daughter strand gap repair (1972); pyrimidine dimer incising endonucleases (1978); adaptive repair pathway (for alkylation damage) (1980–1985); and selective repair of transcribed DNA strands (1987).

Hittelman and Pollard (1982) in their paper on DNA and chromosome repair kinetics, based on the research of γ-irradiation of CHO cells, suggest that "... (1) the rapidly repaired DNA lesions are not important in the repair of chromosomes; (2) only a minority of the DNA lesions measured by DNA elution are responsible for chromosome damage; and (3) chromosome repair might involve more than simply the repair of damaged DNA."

Sasaki (1980) proposed an interesting definition of the role of repair processes in spontaneous and induced mutagenesis for the induction of chromosomal aberrations. In the formation of aberrations spontaneous levels do not directly involve the known DNA repair mechanisms. However, after the exogenous induction of DNA damage, the DNA excision repair capacity is important in the induction of chromosomes aberrations. He also noted that the postreplication step of the repair pathways is possibly involved in the formation of aberrations, while the formation of SCE is associated more probably with replication. Thus, chromosomal aberrations and SCE, according to Sasaki, reflect different damage-tolerating mechanisms and dual-step repair pathways. Repair processes are distinctly defined by the specificity of the acting mutagens and their inhibitors or sensitizers. Wolff (1980) showed that different mutagens induce spectra of lesions with specific proportions that can vary "... in their repairability and the way in which they affect replication."

The modification of repair processes became one of the experimental approaches leading to an understanding of repair mechanisms involved in the clastogenic activity of environmental agents. Wei et al. (1981) demonstrated that the inhibitory activity of chemicals is more directly measured by the decreased levels in cytostatic and clastogenic activity, which reflect the decreased DNA damage, than by the DNA repair synthesis assay. Therefore, the authors emphasized the importance of distinguishing between the inhibition of damage and the inhibition of DNA repair. They noted further that the treatment of cultures by different concentrations of carcinogen (mutagen) is essential. The stage of the cell cycle, i.e., when the inhibitor of DNA repair is added to the culture, is also of great importance, e.g. for investigating the potentiation mechanisms of induced aberrations (Natarajan 1984). The problems pertaining to the modification of mutagenic action through the activation of cell mechanisms, including DNA repair, will be discussed later.

2.2 Induction of Chromosome Aberrations

A very detailed and basic classification of the various types of chromosome aberrations was given by Savage (1976); this will be followed throughout our discussions (Table 3). Besides distinguishing chromosome-type and chromatid-type aberrations, he particularly emphasized the relationships of induced chromosomal structural changes, which he regarded as a result of (1) the highly complex nature of chromosomal organization; (2) the complex biochemical repair mechanisms present in the cells; (3) the great variety of agents which can produce qualitatively identical types of structural changes; and (4) the limitations of microscope observation.

Moreover, more recent experimental studies have been directed toward more practical, quantitative approaches which, in addition, consider more closely the peculiarities of the test systems used also for anticlastogen studies.

One of us (R.A.) based his experiments on knowledge gained from the scientific school of N.P. Bochkov, who preferred practical, quantitative experiments for investigating the formation of cytogenetic changes. As there are not many comparable approaches to the problem of chemical clastogenesis, it seems justified to become acquainted with this type of study in more detail (Bochkov and Chebotarev 1989).

Table 3. Classification of the various types of chromosome mutations (After Savage 1976)

Basic type	Subtypes	Example in metaphase analysis
"Chromosome" type	Interchanges:n	
	Asymmetrical interch.	Dicentrics
	Symmetrical interch.	Reciprocal translocation
	Intrachanges:	
	Interarm intrachanges	Centric rings, pericentric inversion
	Intraarm intrachanges	Acentric rings, paracentric inversion
	Breaks	Terminal deletions, fragments
"Chromatid" type	Interchanges:	
	Symmetrical interch.	Quadriradials
	Asymmetrical interch.	Quadriradials
	Intrachanges:	
	Interarm intrachanges	Centric and dicentric rings, duplication deletion; pericentric inversion
	Intraarm intrachanges	Isochromatid deletions, minutes
	Breaks	Chromatid breaks (deletions)
	Additional aberrations	Triradials, insertion intrachanges, dicentrics with sister union of the acentric fragments
Other types	Achromatic lesions	Chromatid and isochromatid gaps
	Chromosome shattering	Multiple fragmentation
	"Stickiness"	"Sticking" chromosome ends

2.2.1 Chemical Structure as a Basis for Mechanisms of Action

Several types of classification regarding chemical mutagens/clastogens have been discussed so far. The classical grouping of chemical mutagens is based on their major mechanisms of action. Most authors, therefore, discern the following groups of mutagens which, nearly without exception, also act clastogenically: (1) alkylating agents which may also include nitroso compounds and – in a broader sense – some of the substances forming radicals under physiological conditions (the latter, however, are also sometimes regarded as forming a separate group); (2) antimetabolites of nucleic acid biosynthesis; (3) substances that form complexes with the DNA; (4) metals and metalloorganic compounds; and (5) other substances with various, less defined mechanisms of action.

Bochkov and Chebotarev (1989) stated that the relation between the efficiency of a mutagen and its number of reactive groups (e.g., ethyleneimine, chloroethyl, epoxy, methyl groups etc.) is not absolute, as was also shown previously by the experiments of Chang and Klassen (1968) on the clastogenic potency of tri- and hexafunctional derivatives of ethylene imine. Within some of the mutagen groups described above, agents can be found with only one reactive group in their molecule (e.g., monofunctional alkylating agents) or with two or more such groups (bi-, polyfunctional agents). Bochkov and Chebotarev (1989) emphasized that the clastogenic action of a polyfunctional agent might well depend on the distribution of its functional groups on the molecule: they distinguish between "one-centered" substances which bear their functional group(s) at one site of the molecule – connected mainly with one particular atom – and "multicentered" agents, whereby the active groups are situated at different, e.g., opposite, sites of the molecule. In the group of alkylating agents the multicentered agents are able to form so-called cross-links in the DNA by covalently binding two bases of the homologous DNA strands. Within each of these classes the efficacy of the various substances can differ greatly with respect to their ability to induce chromosomal aberrations. Therefore, the clastogenic potency of each mutagen is not only determined by the physicochemical features of its reactive groups, but also by its stability in the environment, the degree of its penetration in the cell, and its ability to reach the genetic material. Of course, all these characteristics may be important in regard to modifiers of clastogenicity. In addition, this fact complicates the interpretation of anticlastogen data.

2.2.2 Dose-Effect Relations

Due to the above-mentioned physicochemical differences in clastogenic chemicals, there are some peculiarities of dose-effect relations which should be briefly mentioned in this context. Dose-response data can also be helpful in understanding the basic principles involved in the formation of chromosomal aberrations. For instance, Chebotarev and Yakovenko (1974) investigated the effect of various concentrations of thiophosphamide on human lymphocyte cultures. They analyzed their data using the following approach:

"Three mathematical models were studied, one of which, being satisfactory for describing the experimental data, was chosen. According to this model a quantity of aberrant cells changes with the concentration of thiophosphamide non-linearly and can be described according to the equation

$$A = \exp\,[-(KC + \alpha)^2],$$

when A is the portion of normal cells, C is a concentration of the mutagen, K and α are coefficients, while exp $[-\alpha]^2$ is the control level of normal metaphases. The quantity of chromosome breaks per cell is described according to the equation

$$X = \exp\,[(KC + \alpha)^2] - 1.$$

The general view of this equation is constant for different phases of the cell cycle."

2.2.3 The Distribution of Chromosome Breaks on Cells

The distribution of chromosome aberrations on the exposed cells can provide valuable information on the mechanism of action by which a respective clastogen displays its activity (Obe 1969; Bochkov et al. 1972; Obe et al. 1972; Yakovenko et al. 1976). It could be shown that the distribution of breaks among exposed cells is described by Pascal's (negative binomial) distribution. "Phosphamide and thiophosphamide inactivation is a one-step process (m = 1) and the inactivation of dipine and photrine is a two-step one (m = 2). This is in good agreement with the peculiarities of these agents' chemical structure: In the first case ethylene imine groups are located near one phosphorus atom, and in the second case they are distributed to different phosphorus atoms" (Bochkov and Chebotarev 1989). This permits the division of the investigated chemicals into "one-centered" and "multicentered" mutagens as described above.

For the multicentered (polyfunctional) mutagens, according to Yakovenko et al. (1974), "the proportion of the aberrant metaphases is described by the following equation:

$$\rho = \frac{C^2}{a + bC^2},$$

while the dependence of the average number of breaks per cell on the mutagen concentration can be described as follows:

$$M = 2 \cdot \frac{1 - \sqrt{1-\rho}}{\sqrt{1-\rho}},$$

where ρ is the proportion of aberrant metaphases, M is the average number of breaks per cell, C is the concentration of the mutagen and a and b are parameters".

2.2.4 Cell-Cycle Specific Action of Clastogens

The classical studies on clastogenesis clearly detected differences in the pattern of the various types of chromosomal aberrations, depending on the phase of the cell cycle irradiated with ionizing radiation, but also depending on the type of chemical clastogen used. In the latter case, S-phase-dependent agents were distinguished from S-phase-independent agents, i.e., chromosome damage by the former group of chemicals could only be detected if a DNA replication took place between the exposure to the mutagen and the analyzed mitosis (Kihlman 1966). In contrast, an S-phase-independent agent was considered to be chromosome damaging during all stages of the cell cycle (Kihlman 1966).

These rather simplified models, however, must be adjusted to real experimental situations. This is particularly valid for anticlastogen studies, as under these experimental conditions the action of the test substances at and on various stages of the cell cycle is of great importance for a correct interpretation of the obtained results.

Bochkov and Chebotarev (1989) discussed the reasons for difficulties in understanding the dependence of the formation of chromosome aberrations on the stages of the cell cycle at which the cells were exposed to respective clastogens. First, there is some mutagen specificity concerning their ability to induce aberrations in certain stages of the cell cycle (see above). Second, the time necessary for their metabolism can vary considerably between different cell types. Third, the specific conditions of the respective experiments can influence the outcome of studies on cell cycle-effect relations. Fourth, the asynchrony of cell populations in most cell culture experiments, particularly in those on human lymphocyte cultures, has to be taken into consideration; and fifth, the action of the respective clastogen on the cell cycle can additionally cause unexpected results. For example, the different sensitivity of human lymphocyte cultures to the action of clastogens has been precisely studied by Bochkov and Yakovenko (1978). They added various ethylene imine derivatives (phosphamide, thiophosphamide, dipin, and photrin) to the cultures at different periods of lymphocyte cultivation. It was found that the "multicentric" clastogens dipin and photrin, if applied during the first cell cycle, resulted in less fluctuation of aberration frequencies than did the "monocentric" agents phosphamide and thiophosphamide. Thus, between both groups of clastogens differences existed even in the shape of exposure time vs effect curves. Their most important finding, however, was that "the maximal sensitivity was observed at 28–32 h of culture, i.e. 24–28 h before harvesting, with the exception of dipin which was equally active from time zero to 32 h. After this maximum, the sensitivity of chromosomes to all the studied mutagens sharply decreased and reached its minimum at 48–52 h".

2.2.5 Cell Proliferation and Clastogenic Action

Cell proliferation is the modifying factor interacting most efficiently with the induction and expression of chromosome aberrations. The classic approach to the analysis of DNA replication in the cells is the labeling of H3-thymidine. However, as an isolated method, it is now rarely used for investigating the changes in proliferation, induced by clastogens, due to the many technical problems arising and the development of new methods of investigation. More recently, BrdUrd-labeling, which is applied for sister chromatid staining (see below), has also been used successfully for discerning cells that passed one, two, or more cycles of DNA replication in the presence of the labeling agent. The estimation of their number allows important conclusions on the peculiarities of proliferation in a given cell sample, but also on its modification.

Cell cycle kinetics can be examined from different points of view and experimental approaches. Obe and Beek (1984) discussed the mitotic selection against the cells with fragile sites. The analysis of micronuclei formation by the cytokinesis-block method for the determination of cytotoxicity and cell cycle kinetics is presented by Krishna et al. (1989) as a preferable alternative to the BrdUrd-labeling technique.

Nevertheless, the application of BrdUrd for the differential staining of sister chromatids is now the most popular method for the differentiation of cell divisions and is a solid methodological basis for estimating the characteristics of cell cycles.

The proliferating rate index (PRI) was given by Lamberti et al. (1983) as:

$$PRI = \frac{1 \cdot M1 + 2 \cdot M2 + 3 \cdot M3}{100}$$

to define cell cycle kinetics. By applying this index, the authors demonstrated the correlation between the increase in PRI and the decrease in SCE levels.

A so-called replication index, which was first used by Schneider and Lewis (1982), shows how many times, on average, a cell has divided in cultivation medium containing BrdUrd.

It is estimated according to the formula:

$$I_r = \sum_{i=1}^{K} iP_i \, , K = 1, 2, 3, \ldots$$

where P_i are the cells of i-division.

Lazutka et al. (1988) presented methods of statistical analysis for replication indices and average generation times. The average generation time (the term was first used by Ivett and Tice in 1982) is equal to

$$T\alpha = \frac{t}{I_r},$$

where t is the time of cell cultivation in medium containing BrdUrd and $T\alpha$ shows how much time one cell needs on average to divide once (Lazutka et al. 1988).

According to an analysis of the average number of cell divisions in a population of dividing lymphocytes, Chebotarev (1985) demonstrated its correlation with the time of fixation of lymphocyte cultures. On the basis of the obtained linear regression equation, he estimated the average time to double the population of dividing lymphocytes to be approximately 38 h; the earliest time that cells enter the cell cycle after their stimulation was estimated to be 8.6 h. It must be noted, however, that within this total population of cultured peripheral lymphocytes there are rapidly proliferating subpopulations which can perform two rounds of replication within a period of 18–24 h, i.e., with a cell cycle time of 12 h or less. This was shown using fractionated BrdUrd-labeling in the laboratory of one of the authors (E.G.; Herrmann, unpubl. data; Gebhart and Kappauf 1980).

The experimental and mathematical models of proliferation processes make data of cytogenetic studies more precise and enable us to study their variability or modification. The fate of induced chromosomal aberrations, for instance, clearly depends on cell proliferation. While stable aberrations can be transmitted to the following cell generation without great difficulties, the instable ones can prohibit a cell from further proliferation or, at least, can cause the formation of micronuclei and their premature chromatin condensation in the following cell mitoses (Beek et al. 1980). Studies on cell proliferation also allow the discrimination of cell subpopulations with different proliferation rates which, in turn, can exhibit various properties which are correlated with the peculiarities of aberration induction. The existence of subpopulations in lymphocytes cultures with different sensitivities to mutagens was shown by Bender and Brewen (1969) for the action of radiation, and by Beek and Obe (1974) for chemical mutagens. Distinct differences were demonstrated in the sensitivity of lymphocyte subpopulations to the clastogenic action of difunctional alkylating agent A139, as the early replicating population of lymphocytes was three times more sensitive than the late replicating one. Moreover, several reasons could be responsible for these differences: "... the sensitivity of human PL to mutagens is by no means unitary, or determined by only one single event, but might depend on the type of primary lesions induced, the type of DNA-repair involved, and, may be, even on some other, non-DNA related, target in the different cell types" (Obe and Beek 1984).

By comparing the data on the peaks of second-division mitosis in cultured human lymphocytes, Wolff et al. (1984) showed the wide range of variation from 48 to 78 h after culture initiation and noted that the variability could be due to differences in the media used for cell cultivation.

Also, the heterogeneity of subpopulations in human lymphocyte cultures is considerable. As reported by Auf der Maur and Berlincourt-Böhni (1979) "... following PHA stimulation 90% of the lymphocytes divide once, about 65% divide for a second and 40% divide for a third time." Since the ranges of cell elimination vary widely, Steffen et al. (1978) presented results which suggest that "under conditions of optimum PHA stimulation about 25–40% of cells die during the first 40 h of culture..." and Yakovenko and Sapacheva (1984) showed that in late fixations (84–108 h) the coefficient of elimination of aberrant metaphases was approximately 0.5.

On the basis of "imitation modeling", the action of kinetics of cell proliferation on the frequency and distribution of chromosome aberrations in cells was investigated. Viktorov (1985) showed that the synchronization of cell culture does not influence the average frequency or the type of distribution of chromosome aberrations in cells in the first division after their formation. The lymphocyte subpopulations can exist mainly due to the difference in the speed of cell division between normal and aberrant cells. This finding seems to be particularly important, because it permits one to neglect the previous asynchrony of the cell cultures in investigations on the frequency and distribution of chromosome aberrations in cell cultures. Cells with changed proliferative characteristics due to aberration induction can be useful tools for models defining the interdependence of both parameters. The mean cycle time in lymphocyte cultures from trisomy 21 patients, which was estimated on the basis of the "transition probability model" for cell cycle kinetics, was the same as in normal cells from control subjects (Morimoto et al. 1984). However, after bleomycin (BLM) treatment trisomic cells showed a marked decline in the aberration frequencies in successive divisions compared with treated control cells. As one of the reasons for that difference the authors suggested the higher levels of dicentrics and rings in the populations of patients with trisomy. "The marked decline of the aberration frequency through mitosis after BLM treatment in trisomic cells also suggests that there was a strong selective loss of more severely damaged X2 and X3+ cells after BLM treatment" (Morimoto et al. 1984). Thus, the levels of aberrations, their types and speed of selective loss may be important for the interindividual variability of clastogenic (and modifying) effects. Steffen et al. (1978) discussed the different responses of thymus-dependent (T) and thymus-independent (non-T) lymphocytes as a possible reason for the heterogeneity of human lymphocyte populations. However, Anderson et al. (1988) argued that "in any case B lymphocytes only account for between 10 and 25% of all peripheral lymphocyte cell populations and, therefore, their contribution to any set of data will usually be minor."

2.3 Induction of Sister Chromatid Exchange (SCE)

Screening of sister chromatid exchanges has become one of the most popular methods used in cytogenetic studies on chemical mutagenesis in the last two decades (Sandberg 1982; Tice and Hollaender 1984).

There are several aspects concerning the similarities and differences in the mechanisms of their induction between SCE and structural chromosome aberrations, but also in their modification, which should be reviewed here before discussing anticlastogenic effects.

An important aspect is the dependence of SCE formation on repair mechanisms. Although there is no unified theory of their molecular formation, all valid hypotheses include repair mechanisms of mainly recombinational nature (see below), which are also assumed to be the target of certain anticlastogenic mechanisms.

A second aspect is the high sensitivity of SCE to various chemical mutagens which allows much lower agent concentrations to be used for its induction than, for instance, are necessary to cause chromosomal aberrations. This fact, at least theoretically, could also be of importance in understanding anticlastogenic action.

Third, indications of a different nature of the mechanisms responsible for SCE and aberration induction provide new approaches to study and classify anticlastogens. The mechanisms by which SCEs are ultimately formed are not as yet generally defined. One of the first observations was the necessity of the primary DNA lesions to pass through a replication cycle to generate SCEs (Wolff et al. 1974). Sasaki (1982) reported the specific dependence of SCE formation on the type, nature, and conformational state of damage and the repair abilities of the cell. All these findings, of course, are also of interest when the modification of SCE induction is considered.

Lamberti et al. (1983) elucidated evidence concerning the dependence of SCE frequency on a number of other factors, e.g., the type of serum or tissue-culture medium used, amount of used BrdUrd, duration of the cell cycle, and fixation time of the cell cultures. Wiencke et al. (1984) presented further information regarding the various physiological, biochemical, and genetic conditions necessary for the induction of SCE.

As regards human lymphocyte cultures, Lindblad and Lambert (1981) demonstrated that although different proliferation rates and SCE frequencies are found in B- and T-lymphocytes studied separately, no correlation could be detected between the percentages of B- and T-lymphocytes in culture and the mentioned parameters.

Emerit (1984), using the example of oxygen metabolites from inflammatory cells, discussed the possibility that active oxygen radicals from endogenous sources could be the origin of a spontaneous chromosomal instability. Increased SCE levels could be found not only in the inflammatory cells from patients with autoimmune diseases, but also in "normal" lymphocytes, if clastogenic factors isolated from the patients' serum were added to lymphocyte cultures of healthy individuals.

Wolff (1978) interpreted the lack of correlation between the SCE levels and chromosome aberrations in cells of patients with classical chromosomal instability syndromes as an argument favoring the different nature of the formation processes of SCE and chromosome aberrations, respectively.

The level of spontaneous SCE in human lymphocytes can vary depending on the culture conditions and on individual factors. A significant decrease, however, can be observed in the baseline frequency of SCE ". . if at least one cell division takes place before the addition of BrdU" (Bochkov et al. 1984). Therefore, it was proposed to compare two periods of BrdUrd exposure using early and late "fixation times" to reveal a realistic picture of the action of environmental mutagenic factors. By this procedure the amount of repair events acting on the SCE lesions in the time lapse between the two fixation points can be estimated.

By comparing their potential to induce SCE and chromosomal aberrations, clastogens can be subdivided into (1) potent inducers of both SCE and chromosomal aberrations; (2) substances readily inducing SCE but not breaks in the same test system; and vice versa (3) potent clastogens with low, or even lacking, SCE-inducing activity (Gebhart 1981). As the potent inducers of both SCE and chromosome damage are most commonly known, in Table 4 examples are presented from groups (2) and (3), which are generally less well known.

Another practical classification was proposed by Lazutka et al. (1988), who found alkylating agents to be the most potent inducers of SCE. There was, however, a ranking of efficiency possible even within this group, from agents that produce cross-links to the much less effective "radiomimetics". The next group

Table 4. Substances differing in their capacities to induce SCE and chromosomal aberrations (Gebhart 1981)

Strong[a] (or marked[b]) clastogens with low[c] or lacking SCE-inducing capacity		Marked[b] SCE inducers with low or lacking clastogenic capacity	
In vitro	In vivo	In vitro	In vivo
Acridine yellow	AAAF	AAAF	Anthracene
Acrylamide	Acrylamide	Acetaldehyde	Benzanthracene
Acriflavine	Patulin	Anilinhydrochloride	Chlorpropamide
Azacytidine		Butylbutanolamine	Chrysene
Benzoate		Butylurea	Dibenzanthracene
Benzo<α>pyrene-syndiolepoxide		Chloroquine	
Bleomycin		Decarbamoylmitomycin C	
Bredinin		Diethylstilbestrol	
Butylbytanolnitrosamine		Dibutylnitrosamine	
Citrinin		Dimethylamine	
Cycloheximide		Fluoresc. brighteners	
Cytosine arabinoside		Glutathione	
Dibutylphthalate		HAAF	
Ethidiumbromide		Methylcholanthrene	
Ethoxycaffeine		Methyldimethylaminobenzene	
Fluorodeoxyuridine		Pyrene	
Fluoruracil		Pyridine	
Ftorafur		Saccharine	
Furazolidon		Sunset yellow	
Hydroxyquinolinesulfate		Trimethylpsoralene/UV	
Hydroxyurea		Urethane	
Mercaptopurine			
Neocarcinostatin			
Nitrosodiphenylamine			
Phenanthrene			
Primidon			
Proflavin			
Propidium iodide			
Quercetin			
Rubber solvent			
Sorbate			
Tiloron			
TPA			
Trisdibrompropylphosphate			

[a]Strong means induced frequency five or more times the control rate.
[b]Marked means induced frequency twice to less than five times the control rate.
[c]Low means induced frequency less than twice the control rate.

also includes directly acting mutagens, i.e., halogenated pyrimidines and DNA-intercalating agents. The third group comprises substances that modulate cellular metabolism, e.g., modifiers of DNA and protein synthesis and some inhibitors of other cellular processes. The last and most complex group contains the agents with less known mechanisms of action, e.g., various drugs, environmental pollutants, etc.

The persistence of SCE depends on the specificity of its inducer. Lambert et al. (1984) demonstrated that ". . most of the SCE-inducing damage caused by treatment with HN_2 and melphalan in G_1 of the first cell cycle is removed before the S-phase of the third cell cycle, whereas damage caused by MMS and adriamycin seems to be more persistent". Littlefield et al. (1984) added: "Although exceptions have been noted, studies of several cell types have shown that a number of different bifunctional agents . . . induce short-lived lesions that are rapidly removed from cycling cells. In contrast, studies of lesion persistence after exposure to monofunctional agents . . . have generally shown that SCE-producing lesions are relatively long-lived in that they persist through several successive cell cycles." These findings were substantiated by previous SCE studies on individuals exposed to cytostatics. Although a clear correlation existed between the individual rates of breakage and SCE, the formation of the latter appeared rather to reflect a long-term effect of therapy on G_0 cells than did the formation of break aberrations (Gebhart et al. 1980). Carrano (1982), in contrast, noted that chronic exposures usually do not cause distinct increases in SCE frequencies in lymphocytes. One of the reasons, in his opinion, is the rapid removal of SCE lesions (within a few days) after each exposure, and only a slight increase after chronic exposure due to "the proportion of lymphocytes with high SCE frequency." Evans (1982a) added to the complexity of these problems by stating that the decrease in the SCE rate is much more rapid than that of chromosome aberrations.

As regards the understanding of the mechanisms responsible for the formation of cytogenetic changes depending on the clastogens' structure and the time of its action, the so-called liquid-holding experiments, i.e., the incubation of unstimulated human lymphocytes (G_0-phase) after treatment with the mutagen, yielded results of particular interest. After treatment with the bifunctional alkylating agent diepoxybutane in this system, a decrease in SCE with the time of liquid holding was noted (Obe et al. 1982b). A similar decrease was also shown for diepoxybutane and methylnitrosourea in a G_1-liquid holding system. However, if the cells were treated with the latter clastogen in a G_0-system, no reduction in the SCE frequency could be detected (Kalweit et al. 1988). To interpret these results, the authors assume a repair of diepoxybutane-induced lesions in both G_0 and G_1 cells and the repair of methylnitrosourea-induced changes which is restricted to G_1 cells.

Contradictory results were obtained when the induction of SCE by the same mutagen was studied in different stages of the cell cycle. Morimoto (1983) investigated the induction of SCE by benzene plus S9-mix in human lymphocyte cultures. In the various experiments with the application of benzene in the G_0 stage of the cell cycle, no induction of SCE could be found, although there was an increase in the SCE level after 40 h of cultivation. The author, therefore, suggested that ". . . the damage can be repaired completely before the cell enters S. . ." This could also explain why SCEs are absent in the lymphocytes obtained from benzene-exposed workers (Watanabe et al. 1980). Morimoto (1984) recommended that the G_1-S boundary in the first of the two consecutive cell cycles before sampling be regarded as the most effective treatment time for human lymphocyte cultures to induce SCE.

The timing of BrdUrd addition seems to be as important as the time of clastogen exposure for a reliable estimate of SCE induction by the latter. Bochkov and Chebotarev (1989) could show that the effect of treating human lymphocyte cultures for 1 h with thio-TEPA, which was followed by the addition of BrdUrd, was unchanged if the addition was performed at up to 36 h of cultivation, but it decreased when the BrdUrd was added at later cultivation times (48, 60, 72 h). The authors explained that difference by the repair of the lesions inducing SCE which arose in the treated cells in the experiments with early BrdUrd addition. In 1975 Perry and Evans reported on SCE induction of CHO cells by various monofunctional and bifunctional alkylating agents, substances producing single-strand breaks in DNA, and by intercalating agents. It was found that the concentrations of mutagens which were necessary for an increase in the SCE rate were much lower than those needed for the induction of structural chromosome aberrations. There were, however, some exceptions to this rule (see Gebhart 1981 for refs.) which led to the assumption that the origin of chromosome aberrations is different from that of SCEs. Therefore, it is now recommended to include tests on chromosomal aberrations and also on SCE in all test systems used in genetic toxicology. It could be shown that those agents which cause SCE much more efficiently, i.e., about 100 times more than chromosome aberrations, induce more aberrations of the chromatid type, while chemicals that induce SCE less effectively (just a few dozen times more or not more than structural aberrations) cause mainly aberrations of the chromosome type (Shcheglova and Chebotarev 1983).

The interpretation of the dependence of SCE induction on clastogen concentration seems to be burdened with less problems than that of induction of chromosomal aberrations. It was demonstrated that for different mutagens the dose-effect curve is linear for SCEs (Shcheglova and Chebotarev 1983). As discussed above, the dose-effect relations for the induction of chromosome damage seem to be more complex.

The distribution of SCEs among cells also differs from that of chromosome aberrations, as the former can be described by a gamma distribution (Yakovenko and Platonova 1979).

All these differences can be taken as further proof for the different nature of SCE and chromosomal aberrations. That means that different mechanisms are also expected for their modification and that differing experimental approaches may be necessary to study their modification.

2.4 Induction of Micronuclei

The estimation of induced micronuclei developed in the last two decades is rapid, inexpensive, and reproducible; thus, it is one of the most popular cytogenetic approaches to analyze the clastogenic effects of chemicals. There is already a tremendous number of publications on this method, and methodological aspects of the test in its classical performance on polychromatic erythrocytes in the bone marrow of exposed animals have been discussed in great detail (e.g., Schmid 1976;

Wild 1978; Heddle et al. 1984; Collaborative study group for the micronucleus assay 1986, 1988; Mac Gregor et al. 1987; Hayashi et al. 1989). Morever, there are also several technical reports on micronucleus studies in other cell systems (see Huber and Bauchinger 1990 for refs.). Therefore, the main possibilities of the method and the essential quantitative results obtained from its application will be considered here.

Micronuclei develop from chromosomes fragments (or chromatids) or from whole chromosomes, which were not included in the daughter cells during mitotic division. The first papers on the micronucleus test appeared for mammalian polychromatic erythrocytes in 1970 (Boller and Schmid) and for human lymphocytes in 1976 (Countryman and Heddle). The method has not change in these years essentially, despite some modifications.

Very important is the universality of the test, i.e., its independence from the karyotype of treated cells or organisms (Matter and Schmid 1971). Besides routine research in the bone marrow of mice and rats, the micronucleus test has been reliably used in the study of clastogenic activity in different objects: hepatocytes of rat liver in vivo (Tates et al. 1980), male germ cells of mammals (Lähdetie and Parvinen 1981; Tates et al. 1983) and man (Lähdetie 1986), peripheral blood of mice (Steinheider et al. 1985) and man (Fenech and Morley 1985), etc.

The induction of micronuclei has been shown for all directly and indirectly acting alkylating agents that display clastogenic activity in vivo, and most of the antimetabolites and antineoplastic antibiotics (Tsuchimoto and Matter 1979). The induction of micronuclei by spindle poisons is of equally high interest. For some chemicals it could be shown that micronuclei originated from induced disturbances and errors in the spindle distribution of chromosomes rather than from structural aberrations (Banduhn and Obe 1985).

Several studies have shown clear correlations between the results of the micronucleus test and other parameters of cytogenetic damage. In the examination of humans exposed mainly to petroleum vapors (Högstedt et al. 1981), such a correlation between the estimates of micronuclei in erythroblasts and polychromatic erythrocytes in human bone marrow and the frequency of chromosomal aberrations in peripheral blood lymphocytes could be demonstrated. Later, in the study of humans exposed to ethylene oxide, Högstedt et al. (1983) showed positive correlations between the increased incidences of chromosome aberrations in lymphocytes and increased levels of micronuclei in polychromatic erythrocytes and erythroblasts of the bone marrow. There was, however, no significant correlation between micronucleus induction and SCE in lymphocytes and the increased frequency of chromosomal aberrations in lymphocytes. Correlations between the levels of micronuclei in the bone marrow and peripheral blood cells were demonstrated also in mice (Steinheider et al. 1985).

With the exception of the above mentioned application of the method to the male germ cells (Lähdetie 1986), there were few important new approaches developed in micronucleus research in man. A method to prepare lymphocytes with preserved cytoplasm (Iskandar 1979) allows for an improved and more precise detection of micronuclei. Additional modifications followed (Huber and Bauchinger 1990). For the analysis of human micronuclei a method proposed by

Fenech and Morley (1985) is now widely applied. In cytokinesis-blocked cells micronucleus frequencies are estimated only in binucleated lymphocytes to diminish artifacts.

A method to examine exfoliated cells of human buccal mucosa was developed by Stich et al. (1982a,b). This approach is also sensitive; its results correlate mainly with different clastogenic properties of environmental mutagens; collection of material is easy and, as will be shown later, this technique also permits the estimation of the effects of biological and environmental modifiers in large groups of individuals (Stich 1986). This method, however, can be applied not only to buccal mucosa, but to various other epithelia, e.g., from the urinary tract (Rosin and German 1985).

To elucidate further the possible modifications of clastogenic action, it is important to analyze the dose-response relationship for the induction of micronuclei. In different test systems, but mainly from studies on mice bone marrow, these dose-response curves can be described routinely by a linear regression, up to high concentrations of chemicals.

A clear dose response for micronucleus induction was shown in V79 Chinese hamster cells for the effect of the direct-acting agent mytomycin C, and for the indirect-acting agent cyclophosphamide (Krishna et al. 1989).

Only a few in vivo results are presented here. The analysis of azathioprine (Imuran) action in the bone marrow of mice and rats has shown a dose-dependent increase in polychromatic erythrocytes with micronuclei (Van Went 1979). Furthermore, the antitumour agents AMSA caused a linear dose response for micronuclei frequencies in the polychromatic erythrocytes of mouse bone marrow (Larripa et al. 1984). In the full-scale micronucleus test of benzo(α)pyrene, the effect in two mouse strains showed a dose dependence of micronuclei in polychromatic erythrocytes (Awogi and Sato 1989). There was a plateau for peak frequencies of micronuclei at high doses of the mutagen. The same assay yielded a dose-dependent increase in micronuclei after exposure of two mouse strains to benzene. In this case, at high doses, the micronucleus frequency, however, decreased (Suzuki et al. 1989). In a report of the third collaborative study by the Collaborative Study Group for the Micronucleus Test and the Mutagenicity Study Group of the Japanese Environmental Mutagen Society on the difference between intraperitoneal and oral gavage application in the micronucleus test (Hayashi et al. 1989), it was stated: "When dose-response curves were expressed as mg/kg, the curves for the i.p. route tended to rise steeply from lower dose levels than did the curves for the p.o. route (e.g., Ara-C, MMS, MMC, COL. VINC). However, when the dose-response curves were expressed as percentage of the LD50 the superiority of the i.p. route generally tended to be reduced (e.g. MMS, PCZ) or even reversed (e.g., DMBA, MMC)."

In the early years of micronucleus research there were many papers on the experimental advantages of the micronucleus test compared to the analysis of chromosome aberrations in the research on dose responses (Frank et al. 1978; Jensen and Nyfors 1979). Now, the realistic estimates are positive, but more moderate. However, many problems remain open. There are not many quantitative models pertaining to dose dependencies to take as a basis for the analysis of modifier action. Of course, there are limitations to the specific sensitivity of the micronucleus test. "... the sensitivity needed to detect clastogenic effects of procarcinogens/promutagens which require tissue specific metabolic activation" (Trzos et al. 1978) is no longer expected. There are also limitations to certain forms of the test during its application in human populations. Sarto et al. (1987) recommended the use of the micronucleus test as an assay in the exfoliated cells of the human buccal mucosa "... for the biological monitoring of groups professionally exposed to very reactive and hydrosoluble substances... " There are contradictory reports on the sensitivity of the micronucleus test in lymphocytes in human populations compared to the chromosome analysis (see Huber and Bauchinger 1990). However, as a comparative new method in mammalian cytogenetics, micronucleus analysis can be developed on the basis of the progress made in methodological approaches.

2.5 Practical Aspects of Clastogenicity and Its Estimation

This discussion of practical aspects of clastogenicity does not intend to give descriptions of techniques or methodological approaches, which can be found in great detail in the respective literature (Hsu 1982; Kilbey et al. 1984). In addition, a great number of reviews deal with the action of chemical mutagens on mammalian or human chromosomes and the definition of the induced aberrations (Gebhart 1970; Buckton and Evans 1973; Evans and O'Riordan 1975; Savage 1976 etc.). Others compare the induction of chromosomal aberrations and SCE (Gebhart 1981; Evans 1982b; Obe and Beek 1982; 1984; Bochkov and Chebotarev 1989).

We will focus on practical aspects, particularly, on the problems of the screening for and the monitoring of clastogenicity which, later on, will also be useful for discussion on the results of anticlastogenesis research.

Reliable scientific conclusions must be based on methodologically correct assays using reproducible standard protocols. A good example of an in vitro assay is the standard protocol for the CHO cell system given by Galloway et al. (1985) which has been evaluated on the basis of so far 106 tested chemicals (Galloway et al. 1987). The recommendations for the evaluation of chromosomal aberrations and SCE are so numerous that a detailed discussion is beyond the scope of this work. However, throughout the pertinent literature, there have been repeatedly seemingly contradictory results which need further consideration. Some could certainly be due differing methodological approaches and protocols, as pointed out by several authors (e.g., most recently by Kirkland 1989). On the other hand, as particularly evident from studies on human lymphocytes, there are also sources of variation which lie in the nature of the target cells used.

One of the controversial problems in the investigation of mutagens in human lymphocyte cultures is the interindividual variability of the sensitivity of these cells, and consequently, of the obtained results. One solution to this problem could be to use the blood of the same donor for the whole series of experiments with one separate agent. However, this is not always possible and, in addition, there are no guarantees that the donor will not undergo some physiological changes, which will thus change the sensitivity of his cells. Also, there are no guarantees that cells of a certain donor are not hypersensitive or hyperresistant to the agents to be investigated. Therefore, many authors recommend the use of blood of more than one donor and "... it is preferable to exclude individuals suffering from virus or other infections or who have received excessive doses of radiation or drugs, or recently received immunization" (Evans and O'Riordan 1975).

The correct planning of the experiment is very important for the reliable conclusions on the mechanisms of the investigated processes. For example, without the necessary quantity of experimental points (at least five) it will be impossible to reveal the real form of the "concentration-effect" curve by its description with mathematical models. To describe the effect of modification, it is sometimes important to induce the clastogenic effect, i.e., modified, in that stage of the cell cycle which is most sensitive to the respective mutagenic action. Information as to whether a mutagen is acting S-dependently or S-independently can be helpful. Using human peripheral lymphocyte cultures, because of the variety of subpopulations with different sensitivities present in individual proportions of the studied blood samples, it will be possible to detect the most sensitive time for concrete culture treatment (under concrete conditions). Preston et al. (1981) recommended also to use more than one continuous treatment time, e.g., with fixations at 3, 8, and 12 h after the treatment has started, to reveal the dependence of aberration induction on different stages of the cell cycle. For more precise information on the cell cycle response, the authors recommended even shorter treatments of 1–2 h.

Very convenient and informative for the work with lymphocytes is the treatment of their cultures in two specific stages, G_0 and G_2. The unstimulated lymphocytes, which were naturally "synchronized" at the G_0 stage of the cell cycle, are an ideal target for treatment by mutagenic agents of noncycling cells, providing a more concrete interpretation of results, if one is interested in analyzing the influence of repair systems. For the research on mechanisms of modification, this possibility is invaluable. The treatment of cells, the majority of which are in the G_2 stage, i.e., before cell division, also has its advantages. In the studies of Wolff and Arutyunyan (1979), it was shown that the alkylating agent thio-TEPA can induce aberrations in G_2 cells, thus confirming the previous data of Chebotarev and Yakovenko (1974). From our findings in N.P. Bochkov's Laboratory, most, if not all, of the investigated alkylating agents can induce aberrations in G_2 lymphocytes. However, the mutagen must be applied in concentrations several (dozens) times higher than that added at other stages of the cell cycle, especially on G_1-S, to obtain the same clastogenic effect, because the G_2 has a low sensitivity to the treatment with aklylating agents. For investigating the mechanisms of clastogenesis, it is also reasonable to use high concentration levels of the respective agent because this permits one to reveal the shape of "time-effect" or "concentration-effect" curves. It is also suitable for the analysis of parameters other than average quantities of aberrant metaphases, e.g., patterns of aberration types, etc. In addition, it may be essential to estimate the distribution patterns of breaks among cells; this can be achieved much better if the average quantity of breaks per cell is higher (Bochkov et al. 1972).

From the existing literature on the use of human peripheral lymphocytes in clastogenesis research, the review by Obe and Beek (1984) can be especially recommended for further information. The authors, among others, not only analyzed the role of repair and cell survival but also the importance of fragile sites. To analyze the sensitivity of early and late proliferating lymphocytes, a series of fixations in the major part of the first mitotic wave, as was recommended for radiation-induced chromosomal aberrations (Bianchi et al. 1982a), can be used.

The differential staining of sister chromatids permits one to "mark" the cells of concrete cell division after the addition of BrdUrd. However, to detect a possible mitotic delay, it is also recommended to use "... a number of fixation times to ensure that cells in their first posttreatment mitosis (in the case of scoring for chromosome aberrations) or second posttreatment mitosis (in the case of SCE scoring) are available for analysis" (Buckton and Evans 1982). Research on fragile sites in human chromosomes (Sutherland 1983), reflecting the changes in the levels of their breakage and reunion, is interesting not only from the viewpoint of population cytogenetics. Possible "fragilization" of chromosomes from special contingents or risk groups can be important for intensified clastogenesis research. The artificial conditions, which cause their expression (changes of folic acid levels in medium, etc.) provide also the opportunity to reveal previously unknown components of the spontaneous breakage or the low level clastogenesis induced by weak mutagens. "Instead of 'induction' one could also speak about 'increase of spontaneously occurring breakage'" (Obe and Beek 1984).

The hierarchy of test systems, i.e., their multi-tier complexes will not be discussed here. This is the subject of genetic toxicology; moreover, besides clastogenesis, all the tests for gene mutations, DNA changes, and others would also have to be included. It should be noted that approximately one-half of such tests are cytogenetic tests and their contribution does not decrease with time. Not only does the analysis of correlations between the cytogenetic and "pure" genetic parameters now play an important role of clastogenesis research, but also the analysis of the modifying products of interaction of a mutagen with the cell and its genetic material. In vitro tests yielding 46% false-positive results, as estimated by Adler et al. (1989), are due partly to artifacts, including changes in osmolarity and pH which can cause toxicity. The authors also provide references on the cytotoxic and cytogenetic effects induced by the physiological state of cells in culture. They recommend regarding those chemicals, which induce chromosomal aberrations only on the basis of their toxic action, as indirect clastogens or agents inducing artifactual effects. As regards the necessity of careful analysis of results, they noted: "The in vitro tests with mammalian cells are not to be regarded as amenable to large-scale screening with a 'cookbook recipe'" (Adler et al. 1989).

Thus, in the study of clastogenic parameters, the investigator, in selecting preferential tests, must know their positive and negative characteristics and their importance in understanding the mechanisms of clastogenesis.

To solve the problem of controlling the levels of clastogenic changes in human populations, it is important to plan research not only in in vitro test systems, but also in man in vivo. Most of the investigations in human cells (more than 90% according to Gebhart 1982) are done on lymphocyte cultures. There is abundant literature on the role of environmental and industrial factors in the induction of chromosomal aberrations (Evans 1982b; Gebhart 1982; Forni 1984; Sorsa 1984), SCE (Hansteen 1982; Sorsa 1984), and micronuclei (Sorsa 1984). Ashby and Richardson (1985) in their paper on the "... tabulation and assessment of 113 human surveillance cytogenetic studies" also presented points of view of different authors on the role of studying chromosomal aberrations in the peripheral blood lymphocytes in exposed populations as "evidence of a clastogenic phenomenon" (Preston et al. 1981), or as "... evidence of a possible carcinogenic/mutagenic hazard" (Harnden 1976), noting the latter (with the similar approach of Purchase 1978) as an "acceptible compromise".

In his analysis of all the then available publications (270) on chromosome aberrations in persons exposed to chemical mutagens, Gebhart (1982) reported that in 20.8% the investigations were performed without a control and in 7.5% this information was not mentioned. Groups consisting of less than ten persons were investigated in 32.5% of all the revised data of those investigations. Bone marrow cells were analyzed in less than 10% of all papers. In only a few papers were data presented on induced aneuploidy or polyploidy.

The toxicological and epidemiological estimates of exposure conditions in the populations or risk groups are very important for planning the studies and reporting the results of any research on monitoring of cytogenetic (and other) parameters. Information on control groups is, of course, obligatory. In her table on the confounding and controllable factors in occupational cytogenetic studies, Sorsa (1984) presented the following exposure conditions: the identification of true exposure and the estimation of the exposed dose; and as sources of individual variability, genetic factors, life-style factors and health. In particular, she noted that smoking should be taken into account in controls and exposed individuals, because, according to several findings reported in the literature, it can increase the levels of SCE, and also (in heavy smokers) of chromosomal aberrations. In more recent reviews the role of smoking and also of alcohol consumption in the increased levels of aberrations and SCE has been reaffirmed (Carrano and Natarajan 1988). Bender et al. (1989) present evidence on the "... substantial effects of age on chromosomal aberration frequencies generally and in SCE frequencies..." The role of sex has not been shown to be important for the levels of chromosomal aberrations, but for SCE. Margolin and Shelby (1985) analyzed literature data and reported that "... women average approximately 0.5 SCE/cell higher than men among normal healthy adults." The results of chromosome analysis in vivo (e.g., for cigarette smokers) can depend also on the cellular immune response that became apparent through the "slower" response of lymphocytes upon their stimulation by PHA (Obe et al. 1982a).

The need for the standardization of conditions of cell cultivation in cytogenetic screening was mentioned previously. However, in cytogenetic population monitoring it is not less but probably even more important, because the conditions of research and the quality of control groups are usually less standardized. Carrano and Natarajan (1988) discussed the role of culture medium and serum in the analysis of chromosomal aberrations and especially the role of BrdUrd concentrations in SCE research, noting, that according to literature data, a tenfold BrdUrd concentration in the cell culture can increase SCE frequencies by 50%.

In most of the reviewed papers the problem of unexposed control groups is discussed in detail. Sorsa et al. (1982) noted that the prospective studies have an "ideal study design" when the blood of individuals prior to exposure can be used as their own control. In other cases Forni (1984) proposed the use of group-mated controls of the same sex and age within ± 5 years, with no known exposure to environmental mutagens. It seems paradox that many authors do not propose the separation of groups into smaller subgroups, while in the investigation on modifier action (as will be shown in Sect. 3.3.2), in contrast, such a subdivision is even recommended. For the discussion on the estimates of the minimal group size, in which the effect of anticlastogens can be best investigated, it is reasonable to present some recommended sample sizes for the estimation of the increase in the chromosome aberration frequencies (Bochkov 1982). Bochkov reported that to estimate an increase of 25%, the minimal number of cells to be analyzed is 38 300 from 63 individuals. To detect an increase of 50 and 100%, these numbers are 10 600 and 3100 from 17 and 5 individuals, respectively (the spontaneous level of cells with aberrations in these estimates is 1.2%). All the presented sample sizes are based on the errors of the first and second kind equal to 0.05. Since many literature sources usually present data on smaller samples, Bochkov (1982) concluded that even a doubling of the levels of chromosome aberrations cannot be detected with sufficient certainty. For a theoretical discussion on sample size, the paper of Archer et al. (1981) is of additional value.

To estimate an increase in SCE, it was recommended to analyze 700 cells from 22 individuals for a 10% increase (if the spontaneous level is 6.5 SCE/cell) or 494 cells from 19 individuals (based on a spontaneous level of 9.8 SCE/cell). For both cases, according to Bochkov (1982), the error is 0.05. If the increase to be detected is higher, the sample size, of course, can be smaller. If the presented estimates are considered, it can be deduced that for the detection of an even smaller increase in the SCE level, logically, a much larger sample size will be needed. Mau (1982) and Whorton et al. (1984) also presented meaningful theoretical considerations on the statistical problems of SCE analysis.

De Jong et al. (1988) discussed the insufficient sensitivity of chromosome analysis for the monitoring of workers who are in contact with only low levels of genotoxicants. Anderson et al. (1981), in addition, emphasized the limited value of SCE analysis after very low chronic exposure to chemical mutagens, also arguing that SCEs might be formed from more short-lived lesions than aberrations. Ashby (1988) suggested that "... techniques which monitor the acute toxicity of a chemical to humans may sometimes be unsuited for the use to predict chronic effects." The possibilities of monitoring by cytogenetic methods are not only limited by statistical protocols that cannot always be fulfilled. In addition, the mechanisms involved in the induction of the cytogenetic changes discussed above can be a serious disadvantage for their use in population monitoring.

By the end of the 1970s, great enthusiasm arose concerning the use of the SCE test because of its seemingly high sensitivity for the action of mutagens and the less time-consuming microscopic procedure of its estimation. After long, intense research, this exuberance has now turned to a more critical view. The reasons for this were discussed in Section 2.3 in great detail. They include, for instance, doubts on the reliability of SCE analysis in population monitoring. But also the scoring of chromosome aberrations in exposed populations is burdened with problems, as this type of damage is largely induced by S-phase-dependent agents which, if acting on G_0 cells in vivo, will allow only a delayed development of aberrations (Evans 1982a,b). On the other hand, other authors stress the high reliability of cytogenetic screening systems: "The analysis of micronuclei or chromosome aberrations in bone marrow cells gives a direct estimate of damage induced in a normally dividing tissue..." (Sorsa et al. 1982). Comparing data from different sources, these authors came to the conclusion that only strong alkylating agents (cytostatic drugs) and ethylene oxide efficiently induce both aberrations and SCE. The reason for the lack of correlation between both parameters in most other population studies is the measurement of different types of genetic damage.

Additional information can be obtained from the analysis of micronuclei, particularly from oral mucosa cells and lymphocytes. The micronucleus method is easier than the analysis of chromosome aberrations and SCEs, and sometimes just as precise. It seems to be quite reasonable to consider large-scale application in groups of individuals at risk as compared to aberrations and SCE. Yet, there are only few studies of this kind. The complimentary character of the micronucleus test to the analysis of cell survival and SCE (Krepinsky and Heddle 1983) does not permit one, so far, to recommend the substitution of the classical aberration analysis by the scoring of micronuclei. Only combined examinations of all mentioned endpoints in the different groups at risk can yield a real basis for a possible preferential selection.

There are some possibilities to study the action of chemical mutagens on somatic cells in real situations. A quantitative approach can be used to estimate the interaction effects of chemical mutagens (Bochkov et al. 1979), and for comparative investigations on mutagenic actions in vivo and in vitro (Bochkov et al. 1986) The latter report presents data on a linear increase in SCE and an exponential increase in chromosomal aberration rates with an increased cyclophosphamide dose in human lymphocytes in vivo and in vitro. The same coincidence of both parameters was also shown in rabbit and monkey lymphocytes, not only with cyclophosphamide, but also with a direct-acting alkylating agent, thiophosphamide (Bochkov et al. 1982; Stukalov and Chebotarev 1983; Bochkov and Chebotarev 1989). The dose was expressed in conditional units representing the metabolite concentration at the time of exposure with due regard to the decrease in metabolite concentration. However, due to the many limitations in our knowledge of clastogenetic mechanisms and the limited possibilities of extrapolation, it is not easy to apply these results to estimates of cytogenetic risk in human populations, One of the most important problems arising now is that of threshold (e.g., Müller 1980; Ehling 1988).

Regarding the problems of extrapolation, one of the most pressing problems still is the extrapolation from somatic to germs cells. However, directing all monitoring of cytogenetic changes to human germ cells is not only difficult, but in most cases impossible. Data from investigations on the action of pollutants on meiotic cells from testicular biopsies are very rare (Hulten et al. 1968), and chromosome studies on sperm cells (Martin 1983; Jenderny and Röhrborn 1987; Zenzes 1987) are not suited for studying larger populations. Regarding the necessity of quantitative risk estimation from germ cells in vivo, Lyon (1982) demanded that this should be performed"... preferably from the same germ cell stage or sex as that involved in the human environmental exposure being considered."

In conclusion, the possibilities of the discussed methods seem to be still much lower in the field of monitoring than they are in the field of experimental screening of mutagenic action. Due to the comparative knowledge of the mechanisms involved in the induction of chromosome aberrations (as compared to SCE) and the higher indicative value of their analysis for the detection of DNA damage and more vulnerable cells (Sorsa 1984), and despite the simultaneous knowledge of all the weak points of this analysis, its important role in cytogenetic monitoring is maintained. It can be supplemented by the analysis of micronuclei and SCE, and the studies can be extended to various tissues. As will be shown below, all these possibilities were utilized in anticlastogen research and, therefore, were discussed here in great detail.

3 Anticlastogens: Data and Problems

3.1 Definition and Classification of Anticlastogens

3.1.1 Definition and Delineation of Anticlastogenesis

An anticlastogen, similar to an antimutagen, in a first approximation, can be defined as an agent which reduces any chromosome-damaging effect resulting from a genotoxic influence in the studied test system. In this sense, an anticlastogenic effect can be estimated by a statistical comparison of the quantitative outcome of experiments performed with the clastogen alone and those including the addition of an anticlastogen. As will be shown below, the high complexity of metabolic reactions and interactions, but also of eukaryotic chromosome structure and specificity, requires the analysis of all possible outcomes in a more sophisticated way, based on the principles of molecular and biochemical mechanisms of the chromosome-damaging action of the clastogen discussed in Chapter 2.

The difficulties imposed upon a clear-cut definition of an anticlastogen are not only elucidated by the complex mechanisms of aberration induction (Chap. 2) but also be specific reactions of several of the protecting agents. For instance, one and the same agent may prove to be an anticlastogen and a coclastogen, depending on the respective experimental conditions. These conditions, among others, include the composition of the used media, the time of application of the presumed anticlastogen in relation to the clastogen (pretreatment, simultaneous, and posttreatment) but also in relation to the phase of the cell cycle to which the substances are applied, and the used cell type or tissue (metabolic capacity), which will be discussed in the following chapters in more detail. All these variables necessitate the standardization of test conditions, especially if comparative conclusions on the efficacy of anticlastogenic action are to be drawn. The various endpoints of modern cytogenetic mutagenicity testing can additionally complicate the definition of anticlastogenic action: which type of clastogenic damage must be diminished by a "real" anticlastogen? In other words, must an agent which significantly reduces the number of induced micronuclei in clastogen-treated cells or tissues, an effect which has been particularly the subject of in vivo studies (see Sect. 3.3.2), also reduce the number of SCEs caused by the same mutagen? Or is such an agent not an anticlastogen, although it considerably reduces the frequency of induced breakage in a given test system and increases the number of chromosomal interchanges, i.e., the aberration type of particular interest with respect to malignancy?

There are several indications from research in this field which clearly document that all these questions are of actual relevance (see, e.g., Gebhart 1972, 1973a,b; Gebhart and Becher 1976). In addition, the definition of the relevant type of aberration may depend on the research interests of the investigator. For instance, in regards to the possible relevance of anticlastogen data to possible anticarcinogenic potency, at the cytogenetic level, the reduction of chromosomal rearrangements by the presumed anticlastogen may be of greater importance than when a general antimutagenic potency is to be defined.

Thus, one practical way of classifying anticlastogens in particular would be to group them according to the type of chromosomal defect which they inhibit preferentially. It is well known that chemical clastogens vary widely with respect to the spectrum of aberrations induced by them (Gebhart 1977). Similarly, anticlastogens differ in their ability to reduce the number of open breaks, of chromosomal rearrangements, and/or of sister chromatid exchanges induced by the mutagens (Fig. 2). In the following chapters, therefore, particular attention will be given to these peculiarities of anticlastogenic activity.

Being very rigorous in definition, an anticlastogen of general importance, therefore, should be a substance which diminishes all types of cytogenetic damage in cells of any type or from any tissue and under all experimental conditions. As will be shown below, only a few substances will meet this very strict definition.

The combined actions of antimutagens also need further consideration. The effects of combined action can be:

1. Additive, when the combined protective effect of two antimutagens (anticlastogens) is equal to the sum of their separate effects;
2. Synergistic, if the combined protective effect of two antimutagens is greater than the additive effect;
3. Independent, if the effect of the combined substances is equal to the effect of one of them;
4. Antagonistic, if the two chemicals together yield an effect that is less than that induced by them separately, or if an inactive chemical decreases the effect of the protective one;
5. Promotional, if a nonprotective agent increases the effect of an antimutagen; or mutually promotional if two agents, nonprotective per se, produce an antimutagenic effect when applied in combination.

It should be emphasized here that the "adaptive response" is not included in our definition of anticlastogenic activity; although due to its eventual effect, it also results in reduced mutagenic effects. Several previous observations of a reduced frequency in chromosomal aberrations following successive exposure to different clastogens of mammalian cells may be subsumed under this action, e.g., the decrease in the number of chromosomal rearrangements induced by N-nitroso-N-methyl-urea in Chinese hamster cells by pretreatment with actinomycin D (Seleznev 1976) or the effect of low doses of sodium arsenite on the number of chromosomal breaks induced by TEPA in another hamster cell line (Rössner 1977). Rieger and coworkers have studied the nature of "adaptive response" in plant cells using a variety of mutagens (Rieger et al. 1982, 1986; Heindorff et al.

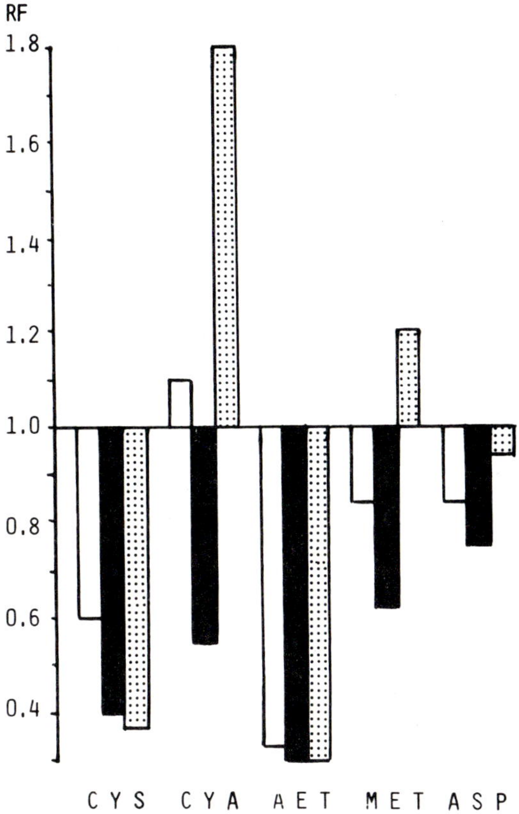

Fig. 2. Differing efficacy of anticlastogens (at their most effective dose) concerning their action on various aberration types (according to Gebhart and Becher 1976). *White columns* chromatid breaks; *black columns* iso chromatid breaks; *dotted columns* interchanges. *CYS*, l-cysteine; *CYA*, cysteamine; *AET*, see abbreviations list; MET, l-methionine ASP, l-asparagine

1987). Besides low clastogen doses they used pretreatment with ammonium chloride, zinc sulfate, or heat shock as inducers of adaptive response. This clastogenic adaptation was found to depend on the agents being used for conditioning and challenging, on the dose of the agents used for pretreatment, and on the time lapse between conditioning and challenging. The authors noted that clastogenic adaptation was observed for S-phase-dependent and S-phase-independent clastogens.

More recently, Wolff et al. (1990a,b), studying the same effect on human lymphocytes in great detail, reported that low doses of X-rays rendered the cell resistant to the induction of chromosomal damage by subsequent high doses of X-rays. If the pretreatment of the cells with low X-ray doses, however, was followed by MMS treatment, a synergistic response was shown. In contrast, the chromosome-damaging effect of MNNG treatment was diminished by preexposure to low doses of X-rays. Similarly, if human lymphocytes pretreated with various concentrations of MMS or MNNG were subsequently challenged with MMS, MNNG, or X-rays, the result varied depending on the agent used for pretreatment:

"When on the other hand, the cells were challenged with MMS, a synergistic response of increased damage was observed. This synergism even occurred if the cells were pretreated with MMS and then challenged with various concentrations of MMS. Pretreatment of the cells with low doses of MMS, however, led to a reduction in the number of aberrations when the cells were challenged with either 1.5 Gy of X-rays or a high dose of MNNG. These experiments show that DNA damage induced by methylating agents is able to induce an adaptive response that makes the cells resistant to induction of aberrations by subsequent exposure to alkylating agents as MNNG or X-rays. On the other hand, irrespective of the agent used to induce the adaptive response, cells challenged with MMS show a synergistic increase in the number of chromatid breaks, indicating that the induced repair system cannot affect certain types of lesions" (Wolff et al. 1990a).

This short excursion into the adaptive response was undertaken for reasons of elucidation and should not divert us too much from our essential aim, i.e., the "real" anticlastogenic effects. Despite the difficulties in defining anticlastogenesis, it is also important to elucidate the ways in which it is effected and the molecular mechanisms contributing to it. Both areas of research led investigators to attempt a more precise classification of anticlastogens.

3.1.2 Classification of Anticlastogens According to the Way They Act

As anticlastogens and antimutagens share many properties, and, in many cases, are even identical, it should be possible to base any classification of anticlastogens on the same general principles used for the classification of antimutagens. As reviewed by de Flora and Ramel (1988) in great detail, several principles can be useful for this classification. For instance, the field of application could be regarded as a basis of classification, thus separating anticlastogens which protect against the mutagenic activity of radiation from those which are active in the induction of mutations by chemicals (radioprotective vs chemoprotective action). However, since antimutagenic activity against chemicals in many cases is caused by the same molecular mechanisms of action as is protection against the mutagenic action of radiation (Gebhart 1974), it seemed to be more appropriate to classify the protective agents according to their mechanism of action and the way they exert their protecting activity.

The ICPEMC expert group on antimutagens and desmutagens in its report (Ramel et al. 1986) used the term "mutagen inhibitors", but noted that the term "antimutagen" is also often used with the same general meaning as the term "anticarcinogen". It subdivided the group of "desmutagens", proposed by Kada et al. (1982) for the inhibitors of mutagenesis, which act outside the cell, into three categories. The first category includes the desmutagens "which directly (chemically and enzymatically) inactivate mutagens." It comprises the inhibitors of formation or uptake of mutagens, acting extracellularly, and includes the extracellular inactivation of promutagens and mutagens. The second category includes desmutagens "which inhibit formation of active forms of mutagens, (a) by inhibition of metabolic activation; (b) by inhibition of formation of mutagens from precursors." The third category includes those desmutagens "which adsorb mutagens by fibers or other high molecular weight substances." More precise is the presentation by the same expert group of the concrete intracellular mechanisms of inhibition: blocking agents and scavengers of radicals via the genotoxic compounds to the reaction sites with cellular targets, then following the agents acting on DNA repair and agents acting suppressively on the route from those targets to the genotoxic manifestation.

A more recently developed classification of inhibitors of mutagenesis and carcinogenesis was presented by de Flora and Ramel (1988): the inhibitors of mutagenesis (clastogenesis), acting extracellularly, are divided by them into three groups: (1) inhibitors of the uptake of mutagens or of their precursors which can act by hindering their penetration or by favoring their removal; (2) inhibitors of the endogenous formation of mutagens, e.g., by inhibiting the nitrosation reaction or modifying the microbial intestinal flora: (3) deactivators of mutagens by physical, chemical or enzymatic reactions. Furthermore, the inhibitors of mutagenesis acting intracellularly are divided into the following groups: (1) modulators of metabolism which inhibit cell replication, or favor sequestration of mutagens in nontarget cells, or which inhibit the activation of promutagens, or induce detoxifying mechanisms; (2) blockers of reactive molecules which act via reactions with electrophils chemically or enzymatically, via scavenging reactive oxygen species, or via protecting nucleophilic sites of DNA; (3) modulators of DNA replication or repair. These agents may increase the fidelity of DNA replication, favor the repair of DNA damage, or inhibit error-prone repair pathways.

In addition, they separately defined anticarcinogens as inhibitors acting on initiated or neoplastic cells and grouped them into modulators of tumor promotion and tumor progression. The first group comprises anticarcinogens which act by inhibiting genotoxic effects, by scavenging free radicals, by inhibiting cell proliferation, by inducing cell differentiation, or by modulating signal transduction. In the second group the authors collected agents which inhibit genotoxic effects (which are also responsible for certain pathways of tumor progression), agents which act on hormones or growth factors, or on the immune system, i.e., physical, chemical or biological antineoplastic agents, or substances which modulate signal transduction.

For each of these groups de Flora and Ramel gave numerous and characteristic examples, several of which will be found in more than one group, thus reflecting the multiple potency of some of the antimutagens (see Table 5).

Most recently, Hartman and Shankel (1990) reviewed mechanisms of antimutagenic and anticarcinogenic action and introduced the term "interception" for mechanisms effecting antimutagenesis by trapping or eliminating mutagens via tissue and cellular organization, and via enzymes or compounds present in the cell or the surrounding fluid. In many instances, these "interceptors" are metabolites

Table 5. Classification of anticlastogen according to their possible mechanisms of action (After the system proposed by De Flora and Ramel 1988)

Desmutagens		
"Uptake inhibitors"	"Inhibitors of endogenous formation of mutagens"	"Deactivators"
Fatty acids: e.g. linolenic acid	Vitamin C Tocopherols Plant phenols Pyridine	Cysteine Cysteamine Homocysteinethiolactone N-acetylcysteine AET Unithiol Cystaphos Gammaphos Vitamin C Vitamin E
Bio-antimutagens		
Modulators of metabolism	Blockers of reactive molecules	Modulators of DNA synthesis (incl. repair)
Cysteine Cysteamine Homocysteinethiolactone AET Unithiol Cystaphos Gammaphos Benzoflavones Naphthoflavones Tannic acid Antipain Aminoharman Phenobarbital Pyridine	Cysteine Cysteamine Homocysteinethiolactone AET Unithiol Cystaphos Gammaphos WR 2721 Vitamin C α-Tocopherol Retinol Resorcinol + Pyrogallol Benzoflavones Naphthoflavones Butylhydroxyanisol Butylhydroxytoluene Superoxide dismutase Catalase Glutathione	Tannic acid Vanillin Cinnamaldehyde Coumarin Ethanol Caffeine Novobiocin WR2921 Interferon?

of low molecular weight. A large series of substances, in this context, has been reviewed by these authors under the aspect of being intercepting antimutagens.

A close interaction of all these ways and molecular mechanisms of antimutagenic action with those which prevent chromosome aberrations builds the bridge between antimutagens and anticlastogens. Most of the mechanisms responsible for the induction of primary DNA lesions, consequently, also cause various types of chromosome damage, and agents acting as antimutagens, therefore, in most cases will also act as anticlastogens.

3.1.3 Classification of Anticlastogens (Antimutagens) According to Their Chemical Reactions

Most of the agents which have been shown to act strongly anticlastogenic from their chemical structure, their metabolism, and their reactivity are certainly not acting according to a single molecular mechanism only. Depending on the site of their reaction, they can interact with various structures and mechanisms causing chromosome aberrations.

In the "early" years of anticlastogen research some concepts on the molecular mechanisms underlying anticlastogenic action were discussed which dealt particularly with radical scavenging, stabilizing enzymes by forming mixed disulfides, increasing the SH-pool of the cell, and other biochemical ways of action (see Gebhart 1974 for refs.). Some of these "historical" but yet valid findings will be discussed in connection with anticlastogenic action against ionizing radiation (Sect. 3.5).

In Table 6 an attempt was made to classify some of the well-known anticlastogens on the basis of their biochemical reaction mechanisms. As can be seen from this table, difficulties arose from the multiplicity of reactions of several of these agents.

Several examples for the involvement of multiple inhibition mechanisms are documented by de Flora and Ramel (1988). For instance, antioxidants, which will be discussed below as efficient anticlastogens, interfere with mutagens by deactivating them extracellularly via chemical reactions, but they also are active intracellularly as reducing agents and radical scavengers. The same is valid for thiols which can react with mutagens or block intracellularly reactive molecules. They also provide an SH-group for stabilizing intracellular enzyme systems and also favor the sequestration of mutagens in nontarget cells. Complex extracellular and intracellular mechanisms of anticlastogenic action are found for vitamins, which on the one hand, act as antioxidants (ascorbic acid, α-tocopherol), but also control

Table 6. Classification of anticlastogens according to their biochemical reaction mechanisms

Antioxidants Radical scavengers	Other antioxidants	Inhibitors of nitrosation	"DNA-complexing" anticlastogens
Cysteine	Superoxide dismutase	Cysteine	Novobiocin?
Cysteamine		Glutathione	Spermidine
AET	Catalase	Vitamin C	Thiol compounds?
Glutathione	Selenite	Vitamin E	
Other thiols	Butylhydroxyanisole	Retinol	
Vitamin C	Butylhydroxytoluene	Plant phenols	
Vitamin E	Ethoxyquine	Pyrogallol	
β-Carotene		Ethanol	
Penicillamine		Butylhydroxyanisole	
		Butylhydroxytoluene	
		Unsaturated fatty acids	
		Cinnamaldehyde	
		Tannic acid	

cell proliferation and inhibit certain forms of mixed-function oxidases involved in the metabolism of xenobiotics, or inhibit binding of clastogens to DNA. Similarly, natural and synthetic sulfhydryl compounds as well as phenols display their anticlastogenic action via their antioxidant properties but also induce various enzyme activities (Ramel et al. 1986; de Flora et al. 1988). In addition, as "physiological" molecules, "bioantimutagens" are naturally involved in normal cellular processes and, besides their "direct" anticlastogenic properties, interfere with various other mechanisms within the living cell. As discussed in detail by de Flora and Ramel (1988), the frequent involvement of multiple inhibition mechanisms and reactions is not only characteristic for a series of antimutagens but can even blur the mechanism most crucial for their antimutagenic properties.

As mentioned above, anticlastogens may not only display their activity separately, but also may interact among another. De Flora and Ramel (1988) presented impressive examples for these interactions, for instance for vitamins and thiol compounds. As will be shown below, these interactions have also been reported from cytogenetic experiments not only to increase but also to eliminate anticlastogenic action.

Cellular metabolism is not only one of the most pressing problems of practical mutagenicity testing, as was reviewed in detail and substantiated by Neis (1986), but also imposes serious problems on antimutagen research. Complex enzyme systems not only activate or deactivate clastogens but also interfere with reactive groups of anticlastogens in order to maintain their activity or to change them into ultimate "active" forms. Under certain experimental conditions (S9-mix) the efficacy of anticlastogens can also be diminished by metabolizing systems (see below). Many further examples of important metabolic pathways involved in antimutagenic action are excellently reviewed by de Flora and Ramel (1988). Before returning to genetic changes at the chromosome level, it must be mentioned that the great variety of biological properties of anticlastogens can also result in adverse effects under certain experimental, intracellular, or tissue-specific conditions.

This was found particularly with several antioxidants which due to their potential as redox systems "either accept or donate electrons, which may alternatively render them either protective or noxious" (de Flora and Ramel 1988 after Stich et al. 1984). According to our knowledge on the formation of mutagenic nitrosamines from nitrates and tertiary amines (Sen et al. 1969), the reaction products of antimutagens and mutagens, if formed at all, in principle could also act detrimentally. De Flora and Ramel (1988) referred to findings, which showed for instance that the reaction products of thiols and MNNG are inactive in an extracellular environment, while they methylate DNA when generated within the cell.

DNA is the target site of clastogenic action, as shown above, and, therefore, the protection of nucleophilic sites of DNA against attack but also an increase in the fidelity of DNA replication, which favors the repair of DNA damage or inhibits error-prone repair pathways, may represent further biochemical mechanisms of anticlastogens. For each of these theoretical possibilities, practical examples have been presented from antimutagen research (see de Flora and Ramel 1988).

3.2 Anticlastogenic Action on in Vitro Systems

3.2.1 Qualitative Data

3.2.1.1 Mammalian Cell Systems

Considering the fact that anticlastogen studies on plant material date back to the 1950s, surprisingly few examinations have been performed thus far on classical mammalian, cytogenetic, in vitro systems, (Table 7), although these are used routinely in genotoxicity testing (Hsu 1982). Several studies, however, have been directed at sister chromatid exchange (SCE) as a genotoxic endpoint and, therefore, will be discussed separately (see Sect. 3.4).

Mainly antioxidants were examined in more detail for their anticlastogenic activity in mammalian cell cultures. In Chinese hamster ovary (CHO) cells, for instance, the addition of 1-ascorbic acid decreased the clastogenic activities of levodopa, noradrenalin, and dopamine, but the respective action of the positive control agent, mitomycin C, was not inhibited by this vitamin (Iwawaki et al. 1988). According to quantitative analysis, using human peripheral blood lymphocyte cultures, these authors could show that the obtained decrease was due to the diminishing action of ascorbic acid on the oxidative, and therefore clastogenic, products of the test substances.

α-Tocopherol, another vitamin with antioxidative properties, inhibited the chemical induction of chromosomal aberrations in Chinese hamster DON cells, if added to the cell cultures 48 h prior to 2,2,6,6-tetramethyl-4-oxo-piperidin-1-oxyl (Lyn-Cook and Patterson 1984). The latter is a free radical that has been shown to induce chromosomal damage in vitro. The authors concluded from their results that vitamin E molecules already present in the cell cultures may react immediately with free radicals, resulting in less free radical damage to macromolecules in the cells ("desmutagenic" action). A similar mechanism could underlie the reduction of benzo<α>pyrene-induced chromosomal aberrations by dl-α-tocopherol (Smalls and Pettersen 1982).

A significant decrease in the frequencies of chromosomal aberrations was also produced by tannic acid in CHO cells treated with mitomycin C, methylmethane sulfonate, and UV, but not with X-rays or bleomycin (Sasaki et al. 1988). These effects of tannic acid were shown to depend on the DNA prereplicational G1 phase of the cell cycle.

The prediction that an inhibitor of induced carcinogenesis may also inhibit chromosomal events related to that process was tested by studying the effects of antipain, a protease inhibitor and known inhibitor of carcinogenesis, on N-methyl-N′-nitro-N-nitroso-guanidine (MNNG)-induced mutagenesis, chromosomal aberrations, SCE, and cell killing in V 79 hamster cells (Kinsella and Radman 1980). Antipain exclusively inhibited MNNG-induced chromosomal exchanges and all other chromosome aberrations. Thus, MNNG-induced DNA lesions lead to chromosomal aberrations which arise through an antipain-sensitive cellular process.

Table 7. Studies on anticlastogenic action in mammalian cell cultures

Anticlastogen	Clastogen(s)	Cell types[a]	Effect[b]	Reference[c]
L-cysteine, glutathione	Ferritin	CHO	+	13
l-Ascorbic acid	Levodopa, dopamine,	CHO	+	1
	Noradrenalin			1
	Mitomycin C		–	1
	Ferritin	CHO	(+)	13
Alpha-tocopherol	TEMPONE[d]	DON	+	2
	Benzo<α>pyrene		+	3
Antipain	MNNG	V79	+	4
Spermine	Adenovirus	CHO?	+	5
Aminoguanidine	Adenovirus	CHO?	+	5
Tannic acid	Mitomycin C, MMS	CHO	+	6
	Bleomycin	CHO	–	6
WR 2721	Trenimon	CHO, IM	+	7
Selenocystine	MNNG	CH	+	8
Na-selenite	MNNG	CH	+	8
Retinol	Aflatoxin B	V79	+	9
	Benzo<α>pyrene	V79	+	9
Indene F 37	Thio-TEPA		+	10
Streptovitacin A	Bleomycin	IM	+	11
"Green tea" polyphenols	Aflatoxin B1	V79	+	12
β-Carotene	MMS, NQO	CHO	+/+	14

[a]Abbreviations: CHO, Chinese hamster ovary cell line; V79, Chinese hamster lung fibroblast cell line; DON, Chinese hamster kidney cell line; IM, Indian muntjak fibroblast cell line.
[b]Anticlastogenic effect; -- no anticlastogenic effect; +/+ anticlastogenic action on different endpoints: chromosomal aberrations and micronuclei.
[c]References: 1, Iwawaki et al. (1988); 2, Lyn-Cook and Patterson (1984); 3, Smalls and Pettersen (1982); 4, Kinsella and Radman (1980); 5, Bellett et al. (1982); 6, Sasaki et al. (1988); 7, Rau and Gebhart (1987); 8, Whiting et al. (1981a); 9, Quin et al. (1985); 10, Kukushkina and Mosse (1975); 11, Tittelbach et al. (1989); 12, Wang et al. (1989); 13, Whiting et al. (1981a); 14, Stich and Dunn (1986).
[d]TEMPONE = 2,2,6,6,-tetramethyl-4-oxo-piperidin-1-oyxl.

Water extracts of green tea and its major constituents, green tea phenols, besides displaying antimutagenic action on *Salmonella* strains treated with benzo<α>pyrene, aflatoxin B1, or 2-aminofluorene and on V 79 cells treated similarly, but without aminofluorene, also decreased the frequency of SCE and chromosomal aberrations in V 79 cells exposed to aflatoxin B1 (Wang et al. 1989). Studies to define the mechanism of the antimutagenic activity of green tea phenols suggest that it may affect carcinogen metabolism, DNA adduct formation, the interaction of ultimate carcinogen, or the scavenging of free radicals.

Antimutagens, however, display their action not only on radiation- or chemically induced clastogenesis, but are also active, for instance, on virus-induced chromosomal changes, as shown, e.g., by Bellett et al. (1982): addition of polyamines protected adenovirus-infected rodent cells from induced chromosomal damage. Spermine was the only individual polyamine which showed this protective characteristic. The diamine oxidase inhibitor aminoguanidine also displayed protective action. The protective effects of both agents, however, were not additive. Maximal protection was obtained when the compounds were added 4.5 h before mitosis, but significant protection was observed up to 1.25 h before mitosis, the latter suggesting an action of the compounds on G_2.

Certainly, future anticlastogen studies must be and will be performed on mammalian cell cultures more frequently than has been done thus far, as these systems, due to their high practicability and easy evaluation, are an important tool in genetic toxicology and, therefore, also in anticlastogen studies.

3.2.1.2 Human Cell Systems

Due to practical rather than fundamental scientific reasons, data on human cells are presented separately here: both of us, for many years, have focused our experiments on anticlastogens from human cell material. Therefore, this field of research presents an obvious opportunity to go into more detail regarding experimental design. In addition, the body of results on anticlastogenicity, obtained by using human cell systems, seems to justify this separate consideration.

Most of the data to be discussed in the following were obtained from in vitro tests, the vast majority of which were performed on cultures of peripheral blood lymphocytes.

3.2.1.2.1 Action on (Induced) Clastogenicity. Although several studies on the anticlastogenic action of sulfhydryl compounds had been performed as early as the 1950s and 1960s on plant material (Avanzi 1957, 1961; Bonati and Savino 1965; Gebhart 1968; Kaul 1969) and on Drosophila (e.g., Röhrborn 1968), the first data obtained from human cell material date back to 1969 when one of us (E.G.) reported on a reduction in the clastogenic activity of myleran (busulphan) in human lymphocyte cultures using l-cysteine. One year later the same anticlastogen was also shown to decrease the chromosome damaging activity of 2,3,5-triethyleneimine-1,4-benzoquinone (= trenimon; Gebhart 1970). The reducing effect was dependent on the protector concentration, on the concentration of the clastogen, and also on the stage of the cell cycle at which the test agents had been applied (Gebhart 1971). Arutyunyan and Kuleshov (1972) presented data on the modifying effect of l-cysteine on the frequency of chromosome aberrations induced by thio-TEPA in human lymphocyte cultures (see Sect. 3.2.2). Arutyunyan (1973), studying the modification of the cytogenetic effects of cytosine arabinoside by l-cysteine, found that "the effect of l-cysteine was more distinct with an increase in the level of induced aberrations. Manifestation of this effect differed at various stages of the cell cycle."

Good protection of human lymphocytes against the chromosome damaging activity of trenimon was also achieved by the addition of cysteamine or β-aminoethylisothiouronium (AET), if clastogen and anticlastogen were added simultaneously to the cell cultures (Becher and Gebhart 1973). The anticlastogenic effect, in this case, again proved to be clearly dose-dependent in these experiments (Fig. 3). The mitotic indices were affected neither by cysteamine nor AET. The practical meaning of these data was supported by the fact that also a commercially available radiation protector exerted an anticlastogenic effect on trenimon-exposed lymphocytes (Stosiek and Gebhart 1974).

In a further study (Gebhart 1973a) the influence of the amino acids l-alanine, l-arginine, l-asparagine, l-glutamic acid, l-histidine, and l-methionine on the clastogenic action of trenimon was tested on the same cell system. An excess of l-asparagine and l-methionine, if added to the cultures simultaneously with the cytostatic, distinctly reduced its chromosome damaging activity. There was, however, no clear correlation between protector dose and protecting efficacy. Both amino acids exerted different effects on the various types of induced aberrations. Studies with 1-methionine revealed different sensitivity patterns of the various cell cycle phases.

The fundamental possibility of protection against the clastogenic activity of alkylating agents also prompted analogous experiments on a nonalkylating

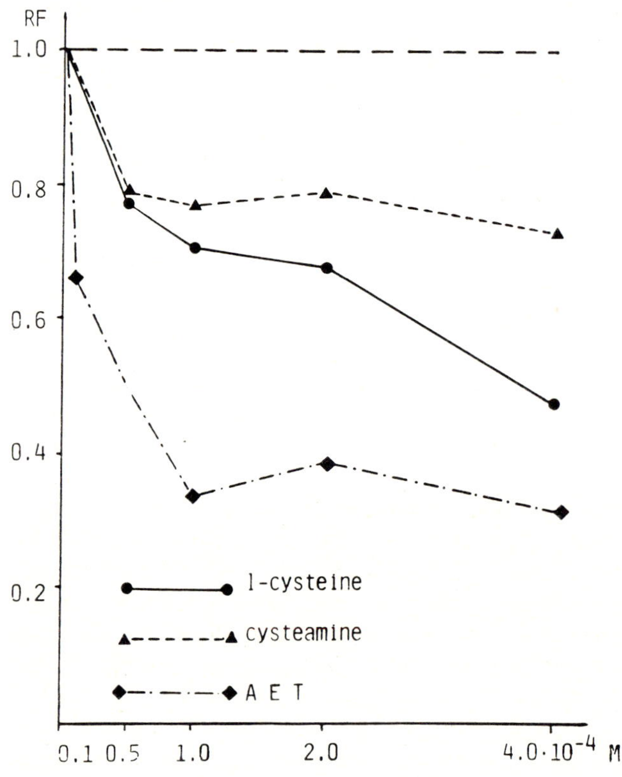

Fig. 3. Summarized dose-effect data obtained from studies with three anticlastogens (Gebhart 1977): simultaneous addition of trenimon (10^{-7} M) and anticlastogens for 24 h prior to fixation. The *RF* (reduction factor) is the quotient of the breakage rate after simultaneous action of clastogen and anticlastogen and the breakage rate induced by the clastogen alone

clastogen, 8-hydroxyquinoline sulfate (8-HQS), which was suggested to exert its damaging effects due to its ability to form chelates. An excess of l-cysteine, cysteamine, and l-asparagine, when added to the cultures simultaneously with 8-HQS, distinctly reduced the clastogenic effect of the latter (Gebhart 1973b). L-glutamic acid and AET showed a less distinct but still significant anticlastogenic effect in these experiments, while the other amino acids studied proved to be ineffective against the 8-HQS.

At the same time Stevenson et al. (1973) reported that the amount of chromosomal (chromatid) damage in human lymphocytes exposed throughout culture to chlorambucil and melphalan was considerably diminished by the presence in these cultures of amylobarbitone. The authors suggested that this could be due to the stimulation of production or activity of "hydroxylase"-type enzymes by the barbiturate and that these alkylating agents, therefore, are more rapidly degraded to derivatives which cause less chromosome damage than the "intact" drugs. Shamberger et al. (1973) found a reduction of the number of chromosomal breaks induced by the carcinogen 7,12-dimethyl<α>benzanthracene by the antioxidants ascorbic acid, butylated hydroxytoluene, sodium selenite, and dl-α-tocopherol. Later, more detailed studies yielded a certain specificity of the anticlastogenic action of the SH-compounds 1-cysteine, cysteamine, and AET as well as of the amide l-asparagine and the amino acid l-methionine (Gebhart and Becher 1976). It manifested itself in specific changes in the spectrum of aberration types induced by the clastogens used (busulphan, trenimon, and hydroxyquionline sulfate), and also in the pattern of intercellular distribution of the induced aberrations. It was clearly dependent on the concentration of the anticlastogens and was also influenced by the nature of the clastogen used. Some quantitative differences in the anticlastogenic activity could also be attributed to differing compositions of the culture media used. The observed patterns of activity of the different anticlastogens specifically differed from those changes in the spectrum of aberrations and patterns of their distribution which would have been induced by a mere reduction of the clastogen dose, as was shown in detail for trenimon. Therefore, a direct reaction between the protectors and the clastogens, in the sense of a desmutagenic effect, could be excluded as the cause of the observed anticlastogenic action.

It is a well-known fact from radiation biology that the combined application of radioprotectors enhances their effects in a hyperadditive way. Thus, attempts were made to confirm this also for chemical clastogens (Gebhart 1978). In screening tests with the combinations cysteine/cysteamine/AET, cysteine/methionine/asparagine, cysteine/serine/homocysteine thiolactone, and AET/homocysteine thiolactone, with trenimon and bleomycin as the clastogens, however, no hyperadditive (synergistic) effect (Fig. 4) could be detected. Sophisticated analyses of the influences on the aberration spectrum exerted by the anticlastogens used, studies on the changes in anticlastogen efficacy by changing the mixing ratios of the applied protectors and on the influence of caffeine on the studied effects, and knowledge of the importance of treatment times, altogether led to the conclusion that anticlastogens manifest their effect in lymphocyte cultures through interaction with the formation process of the primary lesions rather than by influencing repair

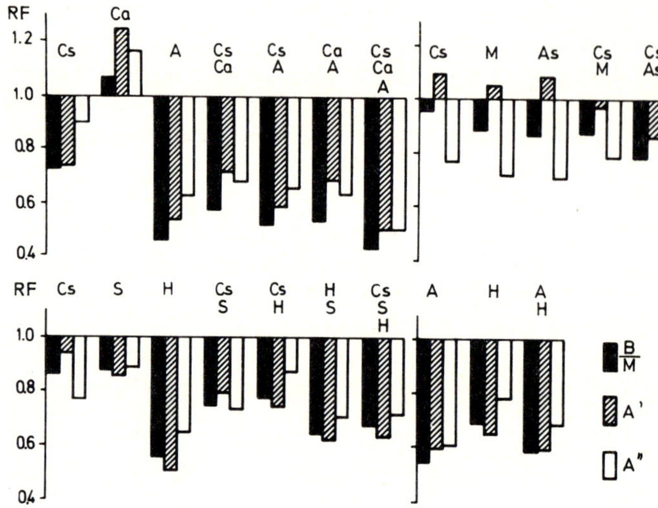

Fig. 4. Anticlastogenic effect of combined application of various protectors. *CS* Cysteine; *Ca* cysteamine; *A* AET; *M* methionine; *AS* asparagine; *S* serine *H* homocysteine thiolactone; *B/M breakage rate; A'* chromatid aberrations; *A* " isochromatid aberrations (Gebhart 1978)

activity. The absence of any synergistic action of the studied anticlastogens indicates a probable common point of action in the cell which is competitively addressed by them. In addition, a mere desmutagenic effect, in the sense of deactivation of trenimon by a reaction with the anticlastogens, was excluded by the finding, that the used anticlastogens practically do not affect the trenimon-induced SCE rate at all (Gebhart 1984). Moreover, several other groups studied the anticlastogenic action on human cell material during the same time (Table 8). For obvious reasons the findings pertaining to anticlastogenic effects induced by "natural" substances merit most interest. A common feature of many of these agents is their antioxidative activity. More recently, it was just this group of antioxidants which attracted particular interest in anticlastogen studies (see Table 8). Not only ascorbic acid or α-tocopherol-acetate actively reduced clastogenic activity (Gebhart et al. 1985), the same was also shown for superoxide dismutase, catalase, butylhydroxyanisol, butylhydroxytoluene (Emerit et al. 1983) and other substances with an antioxidative capacity. However, the weight of these studies shifted to "natural" anticlastogens, as documented in Table 8.

3.2.1.2.2 Action on Spontaneous Fragility. Considering the low spontaneous breakage rate normally found in most cultured cell systems, studies on such a reduction by anticlastogens would impose a heavy burden on the investigator to evaluate large numbers of mitoses for reliable statistical analyses. For this reason, no conclusive study, to our knowledge, has appeared which is specifically dedicated to this problem.

There are, however, some human hereditary diseases which are characterized by a spontaneously increased aberration rate and, therefore, have been termed "chromosome breakage syndromes" (German 1969). These rare autosomal recessive conditions were of basic interest due to the mechanisms underlying their chromosomal instability and their cancer proneness (Arlett and Lehmann 1978; Heddle et al. 1983; Hanawalt and Sarasin 1986). They also offer a welcome opportunity to study the effects of anticlastogens on this spontaneously increased chromosomal fragility. In fact, so far, only a few studies have been concerned with this problem. For instance, Emerit and Cerutti in 1981 could detect clastogenic activity from concentrated ultrafiltrates of media from fibroblast cultures of six patients with Bloom's syndrome by adding them to normal human lymphocyte cultures. The induced frequencies in the chromosomal aberrations depended on the concentration of the ultrafiltrates. Addition of bovine superoxide dismutase to the blood lymphocyte cultures strongly suppressed the clastogenic potency of the ultrafiltrates. From the nature of the obtained ultrafiltrates and on the basis of the anticlastogenic action of the enzyme, the authors speculated that the cells of patients with Bloom's syndrome cannot sufficiently detoxify active oxygen species. Kennedy et al. (1984) achieved a marked reduction in the spontaneous breakage rate in Bloom's syndrome cells by cultivating them in the presence of certain protease inhibitors. They also concluded from their results that the primary defect in this disease may be related to the production of abnormally large amounts of agents, presumably active oxygen species, which are capable of acting like tumor promoters. A partial correction of the dramatically increased levels in chromosome breakage induced by challenging clastogens (e.g., diepoxybutane, butylhydroperoxide, or H_2O_2) in peripheral blood lymphocyte cultures from patients with Fanconi's anemia (FA) and heterozygote carriers of this gene was achieved by simultaneous treatment with various antioxidants (Dallapiccola et al. 1985). A slight effect only, however, could be found in this study of some of the used anticlastogens (e.g., mercaptopropionylgycine, mercaptoethanol, cysteine) with respect to the reduction in the spontaneous breakage rate. In another investigation, niacinamide and adenosine were found to decrease the amount of chromosomal aberrations detected in homozygote and heterozygote lymphocytes of patients with Fanconi's anemia regardless of whether they were treated or untreated with caffeine during G2 phase (Pincheira et al. 1988). The action of the iron chelator desferrioxamine (DFO) on the cytogenetic pattern of cultured lymphocytes from patients with this disease was investigated by Porfirio et al. (1989). The addition of DFO at a final concentration of 10^{-4} M throughout the culture time resulted in a 50% reduction in the spontaneous chromosome breakage of the FA cells. In addition, the clastogenic action of diepoxybutane on FA lymphocytes was also partly counteracted by DFO. The authors believed that these findings support the assumption that one of the mechanisms involved in the pathogenesis of FA might be the impaired capacity of the cells to remove active oxygen species. In contrast, Raj and Heddle (1980) showed that after treatment with superoxide dismutase or catalase, there were no differences in the chromosomal sensitivity between cell strains from Fanconi's anemia patients and the strains obtained from

Table 8. Studies on human cell cultures of anticlastogenic activity

Anticlastogen	Clastogen(s)	Effect[a]	References[b]
L-cysteine	Trenimon, busulphan	+	1–3
	Thio-TEPA, bleomycin	+	4, 5
	Cytosinarabinoside	+	6
	HQS, Mitomycin C	+	2, 3, 7
Cysteamine	Trenimon, HQS	+	2, 3, 8
	Thio-TEPA	+	6
AET	Trenimon, HQS	+	2, 3, 8
	Thio-TEPA	+	6
	Bleomycin	–	9
	Methoxypsoralene/PUVA	+	29
Homocysteine-thiolactone	Trenimon, bleomycin	+	5, 10
N-acetyl-homocysteine-thiolactone	Trenimon, bleomycin	+	5
Reducdyn[c]	Trenimon		5a
N-acetyl-cysteine	Diepoxybutane, bleomycin	+	11
Aminopropylamino-ethylthiophosphate	Thio-TEPA	+	12, 13
Glutathione	Trenimon, (bleomycin)	+(+)	5
Glutathione peroxidase	Phorbolmyristate acetate	+	14
L-methionine	Trenimon	+	2, 3
L-asparagine	Trenimon, HQS	+	2, 3
L-serine	Trenimon	+	2, 10
Sodium fluoride	Trenimon, A 139	+	15
	Trenimon, Cyclophosphamide	+	9
	Bleomycin	–	9
Unithiol	Thio-TEPA	+	6
Gammaphos		+	6
Cystaphos		+	6
Mexamine		+	6
D-penicillamine	Trenimon	+	5
	Bleomycin	–	5
Ascorbic acid	Trenimon, cyclophosphamide	+	16
	DMBA	+	18
	Bleomycin	–+	16, 17
α-Tocopherol	Trenimon, cyclophosphamide	+	16
	DMBA		18
	Bleomycin	–	16
Butylhydroxyanisol	Phorbolmyristate acetate	+	14
Butylhydroxytoluene	DMBA	+	18
	Phorbolmyristate acetate	+	14
Na-selenite	DMBA	+	18
	Phorbolmyristate acetate	+	14
	Na As O_2	+	19
Glutamic acid	HQS	+	3
Superoxide dismutase	Mitomycin C, phorbolmyristate acetate	+	7, 14
Catalase	Mitomycin C	+	7
Mannitol	Phorbolmyristate acetate	+	14
Amylobarbiton	Chlorambucil, Melphalan	+	20
Novobiocin	Cytosine arabinoside	+	21
Hydroxyethylidenediphosphate	Thiophosphamide	+	22
Chloramphenicol	Cyclophosphamide	+	23
Mercaptoethane sulfonic acid	Phosphoramide mustard	+	24
	Acrolein	+	24
	Dimethylhydroperoxycylophosphamide	+	24
WR 2721	Trenimon	+	25
WR 2921	Thio-TEPA	+	26
Interferon	4-Nitroquinoline-1-oxide	+	27
Mercaptopropinylglycine (Thiola)	Benzo<α>pyrene	+	28

healthy individuals. The authors suggested that their results fit the hypothesis of a repair deficiency rather than that of a deficiency of superoxide dismutase as the basic biochemical mechanism of this disease.

It is of particular interest that antioxidants were also shown to reduce spontaneous chromosomal instability, which was attributed to endogenous chromosome breaking factors in certain human diseases. This anticlastogenic action was found for l-cysteine in cells of patients with systemic sclerosis (Emerit et al. 1974), for D-penicillamine and l-cysteine in cells from patients with Crohn's disease (Emerit et al. 1979), and for superoxide dismutase in cells of patients with lupus erythematosus (Emerit et al. 1980).

Further studies on the reduction of spontaneously increased aberration rates must be carried out if these findings, which add to the elucidation of active oxygen species, are to be generalized. The data obtained thus far, however, at least show that cells from patients with these diseases can be used as a valuable model to study not only the biochemistry of the disease but also the mechanisms of anticlastogenic action.

Another aspect, which is closely related to spontaneous chromosomal instability and, therefore, should be briefly discussed here, is the phenomenon of the so-called fragile sites. For the definition and more detailed information on rare, heritable, and common fragile sites the respective vast literature is recommended (e.g., Yunis 1983; Sutherland and Hecht 1985; Le Beau 1986; Sutherland and Simmers 1988).

So far, 18 rare (hereditary) and 72 common fragile sites could be defined, and an association of several of them with breakpoints of chromosomal rearrangements found as specific chromosomal changes in human malignant cells was discussed for some time. Some of these fragile sites are inducible by folate deficiency, two of them by distamycin, one by BrdUrd, but most of them by aphidicolin and caffeine. However, they are also targets of diverse mutagens and carcinogens (Yunis et al. 1987) and have been considered as a possible model for the study of spontaneous chromosome breakage (Jacky et al. 1983).

For our present considerations it seems to be of interest that a protective effect of S-adenosylmethionine against the induction of chromosomal fragile sites was found, although in a rather preliminary study (Zollino and Neri 1987). In this case, according to the assumptions of the authors, it is possible that this methyl donor exerted its protective action through restoration of the correct level of methylation

◀

[a]+ and –, See Table 7.

[b]References: *1*, Gebhart (1969); *2*, Gebhart (1972); *3*, Gebhart (1973a,b); *4*, Arutyunyan and Kuleshov (1972); *5*, Stosiek (unpubl. data); *5a*, Stosiek and Gebhart (1974); *6*, Arutyunyan and Egiazarian (1975); *7*, Raj and Heddle (1980); *8*, Becher and Gebhart (1973); *9*, Gebhart et al. (1984); *10*, Gebhart (1978); *11*, Kartzevski and Gebhart (1987); *12*, Arutyunyan et al. (1977); *13*, Kagramanian et al. (1979); *14*, Emerit et al. (1983); *15*, Obe and Slacik-Erben (1973); Slacik-Erben and Obe (1976); *16*, Gebhart et al. (1985); *17*, Pohl and Reidy (1989); *18*, Shamberger et al. (1973); *19*, Sweins (1983); *20*, Stevenson et al. (1973); *21*, Kishi (1988); *22*, Veltischev and Seleznev (1978); *23*, Vogel and Vrba (1967); *24*, Wilmer et al. (1986); *25*, Rau and Gebhart (1987); *26*, Wolff and Arutyunyan (1979); *27*, Zasukhina et al. (1979); *28*, Galdean et al. (1986); *29*, Abel and Mannschedel (1987).

[c]See text.

of the DNA. Whether or not this study reflects a general modifying capability of the fragile sites by anticlastogens of more general interest must be determined in future research.

3.2.2 Specific Approaches

3.2.2.1 Time-Effect Relations

Since 1970 it was shown that in cultured human lymphocytes a period exists, which is sensitive to protection, i.e., approximately 28–30 h before fixation (Arutyunyan and Kuleshov 1972; see Table 9). The time at which the mutagen was added during the cell cycle made no difference, provided that the anticlastogen was present during that sensitive period. In all cases a decrease in chromosome levels occurred over a wide concentration range of both the anticlastogen and mutagen. As shown in Table 9 and Fig. 5, the action of anticlastogens depends on the time of its addition. The results are consistent with the lack of chemical interaction between the anticlastogen and mutagen and indicate the time period during which the protector must be added, regardless of the time at which the mutagen is added during the cell cycle. The time period sensitive to protection (28–30 h before fixation) corresponds to G_1-S. However, if there had been a chemical interaction between the mutagen and the anticlastogen, then the protocol variants b, d, and e in Fig. 5 would have produced protective effects only when the mutagen and the anticlastogen were added closely together. In contrast, the protective effect would not have been observed with variants a and c. Therefore,

Table 9. The dependence of the protective effect on the time of addition of clastogens and anticlastogens, as revealed by Arutyunyan and Kuleshov (1972). Lymphocytes were harvested after 58–60-h cultivation[a]

Experiment No.	Time of mutagen addition (hour from start of culture)			Time of anticlastogen addition (hour from start of culture)					Protective effect
	0	28	54	0	27	28	29	54	
1		X			X				+
2		X				X			+
3		X					X		+
4	X			X					–
5	X					X			+
6			X					X	–
7			X			X			+

[a]X indicates addition of agents; + protective effect; – no protective effect observed.

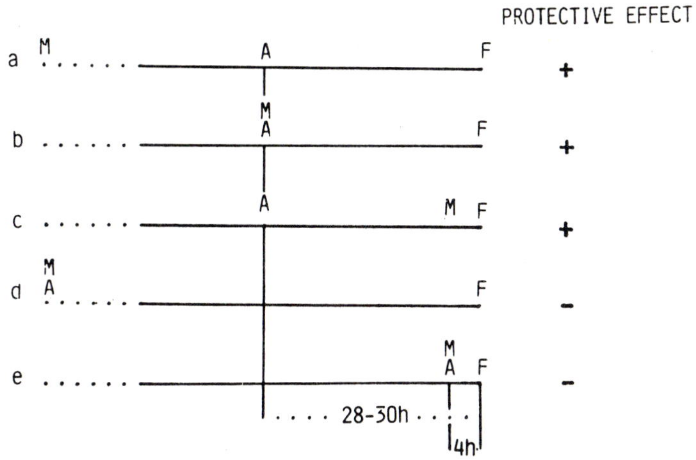

Fig. 5. Experimental protocols for studies on the time-effect relation of anticlastogenic action. *M* Addition of mutagen; *A* addition of anticlastogen; *F* fixation; see text for further details

this test system can be used to determine whether or not there is a true interaction between mutagen and anticlastogen. The most important finding, however, is that the effect of the anticlastogen is maximal when the effect of the mutagen is maximal. The effect of anticlastogen as a function of time is basically the reverse for the effect of clastogen with time. The finding that the protective effect of an anticlastogen is dependent on the time of its addition to the cell cultures, since it must be added during the sensitive period, was confirmed by Slacik Erben and Obe (1976), who studied the protective effect of NaF on the aberrations induced in cultured human lymphocytes by trenimon. "The chromosome analyses show that pre-, simultaneous and post-treatment with NaF suppresses the frequencies of the different types of aberration and leads to a significant reduction of the number of damaged mitoses in cultures treated with trenimon in G_1 as compared with the cultures treated with trenimon alone". The authors had shown a real anticlastogenic activity of NaF. They discussed different mechanisms that could possibly indicate an artifact of the protective effect, but they found no real basis for assuming that this was the case rather than an antimutagenic effect. Furthermore, they reported that "NaF may influence repair processes rather than the induction of cross-links per se." An apparent decrease in the chromosomal aberrations induced by thio-TEPA was caused by the addition of the radioprotective compound S-2-(3-aminopropylamine)ethyl thiophosphate (WR 2721; Wolff and Arutyunyan 1979). For the detection of this protective effect, the "definitely determined division method" was used which utilizes the addition of BrdUrd. Thio-TEPA was added 4 h prior to fixation. The decrease in aberration yield was shown only in rapidly cycling cells, which were in their second division at the time of fixation (Table 10). This was interpreted in the sense that "part of the decrease

Table 10. Aberrations induced in human lymphocytes by thio-TEPA (TT) preceded by 10^{-4} M WR 2721 (WR)[a] (Wolff and Arutyunyan 1979)

Treatment	Total cells		X_1 cells		X_2 cells	
	Aberrant cells (%)	Aberrations per 100 cells	Aberrant cells (%)	Aberrations per 100 cells	X_2 cells Aberrant cells (%)	Aberrations per 100 cells
Control	1.0	1.0	2.0	2.0	0.0	0.0
WR 2721	0.0	0.0	0.0	0.0	0.0	0.0
200 μg/ml TT	4.5	16.5	14.0	15.0	15.0	18.0
WR + 200 μg/ml TT	10.5	12.5	14.0	17.0	7.0	8.0
300 μg/ml TT	27.5	38.5	30.0	42.0	25.0	35.0
WR + 300 μg/ml TT	19.0	22.5	27.0	27.0	11.0	16.0
400 μg/ml TT	33.5	47.0	33.0	53.0	34.0	41.0
WR + 400 μg/ml TT	26.5	32.5	39.0	48.0	14.0	17.0

[a]100 X_1 cells and 100 X_2 cells scored per point.

in aberration yields obtained by the treatment with an antimutagen several hours before the addition of a mutagen is an artifact of cell selection."

Zalinyan (1981, pers. commun.) reported an analogous reduction in chromosomal aberrations by WR 2721 induced by dipine and photrin only in second division cells. Thus, the protective effect in X2 (second division) cells does not depend specifically on the type of mutagen that induces aberrations.

Important evidence of protective mechanisms was obtained from the comparison of time-effect curves from healthy donors and from a family with a xeroderma pigmentosum variant, which was analyzed by Cleaver et al. (1980). Sarkisian (1981, pers. commun.) showed that there were changes in the time-course curves when an anticlastogen was present in the cultures from healthy donors but no changes appeared in the cultures from xeroderma patients from the mentioned family. One of the reasons could be the absence of some metabolic processes that are normally activated by protectors in the cells of xeroderma patients.

One of the most interesting anticlastogens revealed in the 1980s is natural interferon (IF) (Kirkova et al. 1980; Zasukhina et al. 1981; Suzuki 1986; Makedonov et al. 1989). It was shown that natural IF (α) protects chromosomes from chemical mutagens and radiation action at both the level of chromosomal aberrations and SCE due to the activity of some repair processes (Makedonov et al. 1989). Recombinant IF (α2) as an anticlastogen (Sarkisian, Movsesyan, and Zasukhina 1989, pers. commun.) at a concentration of 37 IU/ml produced a significant protective effect on the aberrations of chromosomes induced by thio-TEPA and photrin. More detailed results are the following: thio-TEPA, if added 28 h prior to fixation at a concentration of 1×10^{-5} M induced 14% of the aberrant metaphases. However, with the addition of IF 28 h prior to fixation, the level of aberrant metaphases decreased to 6% (<0.05). In the experimental variant with a thio-TEPA concentration of 2×10^{-5} M, the addition of IF decreased the level of aberrant cells from 20 to 8% ($p < 0.01$). With the increase in concentration of thio-TEPA to 3×10^{-5} M, the level of aberrant cells increased to 26%, while the addition of IF decreased it to 12% ($p < 0.5$).

The analogous protective effect was shown for the action of IF at the levels of aberrations induced with the "multicentered" alkylating agent photrin. Photrin at a concentration of 0.25×10^{-5} M induced a 12% rate of aberrant metaphases, which was decreased by IF to 6% ($p < 0.05$). A concentration of photrin 0.5×10^{-5} M induced 21% aberrant metaphases and in combination with IF this rate decreased to 12% ($p < 0.01$). In the cultures treated with photrin at a concentration of 1×10^{-5} M, 30% of the aberrant metaphases was induced and the addition of IF diminished this level to 12% ($p < 0.01$). Thus, it can be seen that during the action of IF (α2) in our test system, the same effects were revealed, which were shown above for thiols. This confirms the low specificity of the protective effect concerning the chemical type of anticlastogens. Of equal importance is the action of IF in the mentioned sensitive period of the cell cycle. It is of interest to note that in the same series of experiments, recombinant IF (α2) did not decrease the levels of SCE induced by thio-TEPA and photrin. These data confirm the results of Yakubovskaya et al. (1988) on the differing protective effects of natural leukocyte IF (97) and recombinant IF (α2), although both of them stimulate cell proliferation. However, at the level of chromosomal aberrations in our test systems, as can be seen, recombinant IF displayed pronounced protective activity, analogous to the natural leukocyte IF. This was also shown by another series of experiments (Table 11). To understand the potential of our test systems, the finding was very interesting that some modifiers of mutagenesis, which are known sensitizers of mutagenesis and mainly increase the effect of some chemical mutagens, can unfold their activity in the same periods of cell cycle as anticlastogens. This research was done, in part, in the Laboratory of Cytogenetics of Yerevan University (Arutyunyan and Zalinyan). For instance, the action of caffeine was investigated with respect to the sensibilization of the chromosome versus the effect of gibberellic acid, a plant growth regulator. In all investigated concentrations gibberellic acid produced weak clastogenic activity in human lymphocytes. Its maximal concentration (3×10^{-3} M) induced 4.3% aberrant metaphases. However, the addition of caffeine at a final concentration of 10^{-3} M after 29-h cultivation increased the level of aberrant metaphases to 17.3%. The average frequency of breaks per 100 cells increased from 4.3 to 21.3 (Arutyunyan and Zalinyan 1987). Later, it was demonstrated that the maximal potentiation of the chromosome-damaging effect of gibberellic acid is achieved when caffeine is added 30 h before fixation (at a total culture time of 76 h). The increase in aberrations, however, was much lower when caffeine was applied only during the last 4 h before harvesting (Fig. 6). Chromatid exchanges were mostly observed in the series of experiments in which caffeine was added to the cultures after 29 or 46 h of cultivation. Analogous results were obtained for another modifier of repair processes, 3-aminobenzamide, which increased the level of chromosome aberrations induced by gibberellic acid two- to threefold. This increase was observed by the authors

Table 11. Effect of recombinant interferon (IF) on lymphocyte cultures treated with photrin. A, B, and C are donors of blood samples for the experiments (TF Sarkisyan, ES Movsesyan and GD Zasukhina 1989, pers. commun.)

Experiment	Donors	Aberrant cells (%)	Total breaks per 100 cells	Single breaks	Double breaks
Control	A	2	2	0	2
	B	2	2	1	1
	C	2	2	1	1
IF	A	5	5	3	2
	B	5	5	2	3
	C	4	4	1	3
10–5 M	A	14	14	6	8
	B	12	12	3	9
	C	12	12	6	6
+ IF	A	6	6	2	4
	B	5	5	2	3
	C	6	6	3	3
2.10–5 M	A	20	20	12	8
	B	19	19	9	10
	C	19	21	12	9
+ IF	A	8	8	5	3
	B	5	5	4	1
	C	5	5	3	2
3.10–5 M	A	26	26	15	11
	B	29	30	11	19
	C	27	27	7	5
+ IF	A	12	12	7	5
	B	22	22	5	6
	C	15	15	7	8

in the integral mix of cell subpopulations. However, when differential staining of sister chromatids by BrdUrd was used to label the respective cells, the effect of 3-aminobenzamide was mainly evident with X1 cells. In these experiments the modifier was added to the cultures at 29 and 46 h, and the clastogen at 46 h of cultivation.

The effect of the noted modifiers seemed to support the evidence that the sensitive periods of their action are universal. Although extensive research on these modifiers of mutagenesis has been carried out (caffeine, Kihlman 1977; 3-aminobenzamide, Schwartz et al. 1985) and is still in progress in several laboratories, the relationship between their efficiency and repair processes has yet to be elucidated.

Although this may seem paradox, evidence for the anticlastogenic potential for cell systems can also be revealed by studying bidirectional processes, as shown for the coclastogens, caffeine and aminobenzamide. On the basis of the presented results, the conditions necessary for a realistic study of anticlastogenic effects include the following aspects. The anticlastogens have to be added at an effective concentration during that period of the cell cycle which is most sensitive to them.

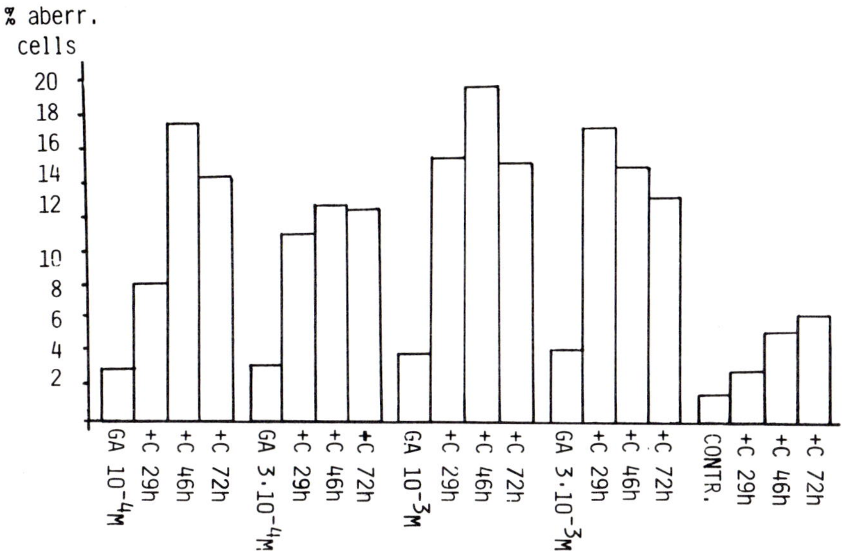

Fig. 6. Effect of various caffeine (*C*) treatments (29, 46, 72 h, time of addition to the cultures) on the clastogenic action of various doses of gibberellic acid (*GA*) (After Zalinyan, pers. commun.)

The level of aberrations induced by the clastogen, however, has to be sufficiently high to reliably detect the protective effect. The latter parameter may be composed of effects of various cell populations, since the added anticlastogen may act differently on cell subpopulations. It cannot be excluded that it could also be a result of "desmutagenic", or with respect to our discussion, "desclastogenic" effects produced by a direct interaction of clastogen and anticlastogen. However, it can also be an artifact in the sense that test substances induced changes in the speed of proliferation of cell subpopulations with different responses to the mutagen or anticlastogen.

On the basis of the presented data regarding time-effect relations, we can model the situations which could create erroneous conclusions on the mechanisms of the revealed protective effects. If, according to experience from radiation genetics, the anticlastogens are added to the cell cultures with the clastogen simultaneously, and also e.g., 1 h before or after the mutagen treatment, we could obtain the following. Because the mutagen can be added at that point of the cell cycle which is most sensitive to it (e.g., for S-phase-dependent agents this is, of course, the S-phase) and because the addition of the anticlastogen coincides with the G1/S borderline, a protective effect can be demonstrated when both agents are added simultaneously. However, by increasing the interval between anticlastogen application prior to or after the addition of the clastogen, the observed protective effect appears to diminish. Thus, erroneously, the conclusion could be drawn that the anticlastogen exhibits its protective effect optimally if added simultaneously with the mutagen. However, within the cell cycle, the interval between clastogen and protector application does not seem to be as important as the addition of the anticlastogen at that period of cultivation which is most sensitive to it.

Research on different anticlastogens has shown that this sensitive period does not always depend on the type of the mutagen used. In addition to using alkylating agents in these experiments, cytosine arabinoside, a pyrimidine analog, was also used, whereby the same dependence of the protective effect of the studied anticlastogens on the culture period of their addition was demonstrated.

Another important point is the observed different sensitivity of cells that have passed one or two replication cycles between their exposure and the moment of harvesting (Wolff and Arutyunyan 1979). One possible explanation for the restriction of the protective effect to X2 cells (i.e., cells in their second division after application of the test substances), besides the partial artifact generated by differing speeds of proliferation of the exposed cell subpopulations, may be as follows: it is likely that WR 2721, which was added to the cultures at 22 h of cultivation, activated the cell systems (regardless of their nature) in the "fast" subpopulation of cells that could complete two cell divisions before their fixation, i.e., the X2 cells. However, 22 h of cell culture was not the sensitive period for anticlastogenic action on the "slow" cell subpopulation, and therefore, no protective effect could be found in those X1 cells. Thus, regardless of when the substances were added in relation to the time of culture fixation, a proportion of cells exists which are more or less sensitive to the action of anticlastogens (or any other modifiers). In every experiment investigating the protective mechanisms, this proportion must be taken into account. However, the possibility of including artifact effects must always be considered.

The main processes that are connected with the time of addition of anticlastogens, i.e., the sensitivity of average cell cultures to the action of a modifier and the sensitivity of cell subpopulations within cultures, are very important for the investigation of anticlastogenic effects at optimal points of the cell cycle. If the "average" estimates of effects are based on a few experimental points in the proliferation cycle or on a few unidentified subpopulations, changed or artifactual protective effects can be revealed. Thus, the main result of all the attempts discussed above is that a time localization of protective efficiency could be determined and a methological approach could be established which permits decisions on whether an observed effect is real or artifactual. Without such an approach, the investigation of any modifying agents would be incomplete.

3.2.2.2 Concentration-Effect Relations

The action of a few radioprotectors on different concentrations of thio-TEPA was investigated by Kagramanyan et al. (1979). The protectors included the mentioned WR 2721; S-3-(3-aminopropylamine)ethylthiophosphate (APAETP-3,3), 2-aminoethylthiophosphate sodium salt (cystaphos). The concentration of the protectors varied from 10^{-4} to 2×10^{-4} M, the temperature of incubation of mutagen and protectors (after 28-h cultivation and 1-h exposure, followed by washing and the addition of new culture media) was 37 °C; in one of the experimental series the incubation temperature was 42 °C.

The equations given by Chebotarev and Yakovenko (1974) for the interpretation of our concentration-effect curves for thio-TEPA with and without the protector were used. This was important because most previous papers dealing with the modification of mutagenesis compared only single concentrations of mutagens and their effects with and without protector. The authors proposed a comparison of the angles of inclination of regression lines representing the dose effect of cytogenetic changes with and without anticlastogen as the measure of modification (protector or sensitization). The linear index of modification (LIM) can be used:

$$\text{LIM} = \frac{\text{K} - \text{KM}}{\text{K}} \cdot 100\%,$$

where K is the slope of the regression line without modifier, and K_M is the slope with modifier. If the effect is not protective, but sensitizing, we will obtain negative estimates of LIM (Arutyunyan 1985).

The regression lines based on the experimental data from Kagramanyan et al. (1979) are presented in Fig. 7. As can be seen, WR 2721 had the most consistent protective effect in two out of three series of experiments, and was used as a reliable standard protector in further experiments. The estimates of K and LIM for the levels of aberrant metaphases in the three treatment modes on the basis of our experimental data are presented in Table 12. The main effect of the protectors was a decrease in single breaks; they were less effective in diminishing the number of double breaks and the level of exchanges. These results agree with most data on the action of anticlastogens (see, e.g., Gebhart 1973a,b) except perhaps the results obtained with the radioprotective drug Reducdyn (Stosiek and Gebhart 1974), as this also reduced the number of chromatid interchanges induced by trenimon. The authors reported that Reducdyn actually inhibited primary breakage instead of being involved in stabilizing enzymatic reunion (repair?) systems, as usually suggested in similar cases.

In a few series of experiments (R.A.) the concentration dependence of the protector effect was investigated. Five different thiols and aminothiols exerted a protective effect in concentration ranges of 5×10^{-5} to 2×10^{-4} M; there was no linear dependence of the protective effect on the concentration of the protector. This can have several reasons. Obviously, anticlastogens are not neutral to different cell processes. Moreover, it is not unusual that under some conditions and in some concentrations they can demonstrate even some clastogenic or cytotoxic effects. Becher and Gebhart (1973) showed that cysteamine in comparatively low concentrations increased the levels of chromosomal aberrations and the incidence of tetraploid cells in human lymphocyte cultures. This was also shown for the investigated dose 4×10^{-4} M, which was comparable to therapeutic doses. This concentration also practically doubled the level of the mitotic index. In the same study no linear dependence of the anticlastogenic effect from the investigated concentrations of cysteamine and AET (β-aminoethylisothiouronium) was shown.

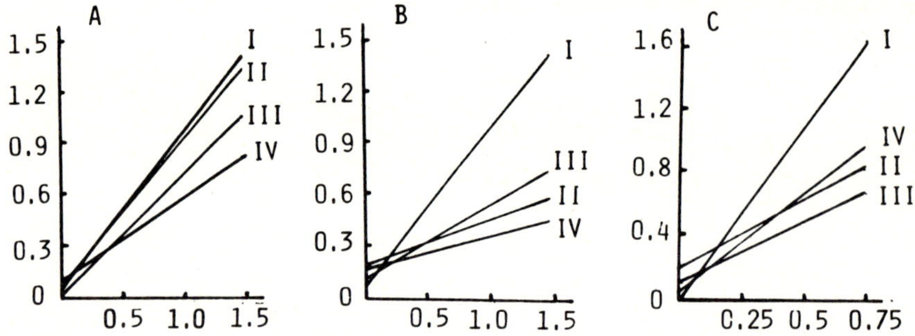

Fig. 7A-C. The dependence of average chromosome breakage levels (ordinates) vs thio-TEPA concentrations (abscissa) for three different treatments. **A** Concentration of the anticlastogens 10^{-4} M, incubation temperature 37 °C. **B** Concentration of the anticlastogens 2×10^{-4} M, incubation temperature 37 °C; **C** concentration of the anticlastogens 10^{-4}M, incubation temperature 42 °C. *I* Thio-TEPA, *II* thio-TEPA + WR 2721; *III* thio-TEPA + APAETP-3,3; *IV* thio-TEPA + cystaphos. (After Kagramanyan et al. 1979)

Table 12. Statistical result of experimental data on percentage of aberrant metaphases and average breakage rates (number of breaks per 100 cells) (After Kagramanian et al. 1979)

Experiment	Temperature of Incubation	For aberrant metaphases		For average breaks/100 cells	
	°C	K	LIM%	K	Lim %
Thio-TEPA	37	0.039	–	0.043	–
+ WR 2721 (10–4 M)	37	0.037	5.30	0.039	10.30
+ APAETP 3,3 (")	37	0.039	0.46	0.041	6.06
+ Cystaphos (")	37	0.024	38.63	0.026	39.59
+ WR 2721 (2.10–4 M)	37	0.011	72.05	0.012	71.41
+ APAETP 3,3 (")	37	0.20	49.21	0.019	56.88
+ Cystaphos (")	37	0.009	78.22	0.009	79.17
Thio-TEPA	42	0.087	–	0.100	–
+ WR 2721 (10–4 M)	42	0.039	55.28	0.040	60.54
+ APAETP 3,3 (")	42	0.032	63.01	0.034	65.91
+ Cystaphos	42	0.053	38.97	0.056	44.53

This finding agrees with most results in anticlastogen research on cell cultures. For example, data of Emerit et al. (1983) on the effect of antioxidants on phorbolmyristate acetate-induced chromosome aberrations in human lymphocytes showed a decrease in the levels of aberrations by glutathione peroxidase of approximately 50% in a wide range of concentrations, 0.06 to 0.30 μM/ml.

To analyze the data of Arutyunyan and coworkers, the correlation matrix, based on the results of a series of experiments, was used to determine the most important parameter relationships and the levels of deviation from linearity of the regression curves. Furthermore, cluster analysis was applied, thus providing the possibility to attribute the objects on the basis of their similarity to one or different clusters.

With the obtained correlation matrix dendrograms were established, i.e., diagrams with branches showing the relationships between the characters. The first dendrogram, based on the results of our data (R.A. with Zalinyan) is presented in Fig. 8A. Thus, 144 experimental variants (dipin and photrin without anticlastogens, and with the addition of WR 2721 and cystaphos) were analyzed. As can be seen from the dendrogram, the cluster for the type of anticlastogen and its concentration ($r = 0.859$) is related to the cluster of cytogenetic characteristics (quantity of aberrant metaphases and breaks), single and double fragments (all per 100 cells) at the level of $r = 0.528$. This whole cluster, but showing less relationship, correlates with the cluster representing the type of mutagen and its concentration ($r = 0.123$). The second dendrogram is based on the results of another series of experiments with 104 variants (Fig. 8B), but here only the variants of combined treatment with the mutagens and anticlastogens mentioned above were analyzed. In this case the relationship between the type and concentration of mutagen and the cytogenetic characteristics is more obvious ($r = 0.210$). This cluster is related to the cluster describing the type of anticlastogen at the level $r = 0.132$.

The results of cluster analysis thus demonstrated the relationships of the cytogenetic parameters with the characteristics of the anticlastogen and clastogen. This approach can be applied in the investigation of many parameters involved in anticlastogenic research in order to understand their hierarchy. However, regression analysis is a better tool to reveal the quantitative aspects of the investigated processes. For example, in the same experiments by applying stepwise regression analysis, mathematical models were obtained that describe the effects of anticlastogens. One of them describes how the frequency of aberrant metaphases depends on the treatment of cultures by dipin and the anticlastogen WR 2721, i.e.:

$$E = 1 - \exp[-(0.357 - 0.430\, C_A + 0.141\, CM^2 - 2.612 \times 10^{-3}\, CM^2 - 6.771 \times 10^{-3}\, C_M C_A + 5.611 \times 10^{-3}\, C_M + 5.501 \times 10^{-5}\, C_M^{1.5})].$$

Here, E is the effect, CM the concentration of mutagen, and CA the concentration of anticlastogen. With this model it was possible to predict the obtained effect of each desired concentration of mutagen and anticlastogen.

The mathematical model of action of mutagen and anticlastogen is represented by the function $R(t, r)$, where R is the velocity of change in the amount of aberrant cells depending on time t, and r is the fraction of cells affected by the protector (Danielian et al. 1978). The limit is equal to $\lim R(t, r) > 0$, where $t \to \infty$. This means that independently of the quantity of protector molecules, the frequency of chromosome aberrations in the cells remains greater than 0. Thus, the model is quite realistic, since this reflects that actually observed.

The second question related to our mathematical model was whether the velocity of change of $R(t, r)$ depends on the frequency of chromosome aberrations in those variants treated only with mutagen. This question is very important because in radiation cytogenetics, there are many explanations for the decrease in the radioprotective effect on the low levels of aberrations. Moreover, it is not clear whether this is due to the fact that there is a real threshold for the protective effect, or whether, due to statistical reasons, very large samples are necessary to investigate the low levels of aberrations. A similar problem occurs also in the application of protectors in chemical mutagenesis. The analysis of our model showed that if $R_0^{(2)} > R_0^{(1)} > 0$, then R(t, r), beginning from some $t > 0$, decreases more intensely if the level of aberrations is higher. Therefore, the application of this simple model, without considering repair and other processes, fits the experimental data.

The use of distribution analysis to investigate the mechanisms of protectors is limited, because in the experimental variants with additional protector, a high level of breaks per cell is not usually observed, thus it can be described by many distributions. However, in our experiments it was described by both Poisson and Pascal distributions. If the protector induces large distortions in the lymphocyte cell cycle, then the empirical distribution will not be described by a Pascal distribution (Table 13).

The interpretation of the relations obtained in the analysis of protective effects can be discussed from the viewpoint of possible mechanisms rather than from practical conclusions (see Chap. 4).

One group of researchers (R.A.) based its investigations of anticlastogens on the quantitative data from clastogenesis research. The linear index of modification (LIM) was based on the "linearization" of concentration-effect curves according to the approach of Chebotarev and Yakovenko (1974). This index permits the comparison of data from different anticlastogens effects which were obtained in various previously incomparable concentrations of mutagens. Moreover, one of the very important conclusions of the mentioned paper was that the chromosome reacts with thio-TEPA as a two-stranded structure. By using the same equation for the "linearization" of the concentration-effect dependencies, it was possible in all series (including the analysis of data from cultures treated with mutagens and anticlastogens) to draw the following conclusion: In the variants with additional anticlastogen, the character of the interaction of mutagen with chromosome is unchanged, as shown for all the investigated alkylating agents, i.e., the reaction with functionally "two-stranded" chromosomes.

Different statistical approaches to analyze the same experimental data revealed, at various levels, the same relationship between the investigated parameters. The most important relationship demonstrated was that between the type and concentration of anticlastogen and the concentration of the mutagen. However, generally no particular specificity in the dependence of the effect on the type of mutagen was shown, even for the modified effects of largely differing "one-centered" and "multicentered" alkylating agents.

The obtained effect of anticlastogens was not so distinctly expressed in the variants with comparatively low and high levels of clastogenic changes, induced by mutagens. This is perhaps the most universal drawback for many results on anticlastogenesis reported by different authors. We do not believe that the insufficient sample sizes are the only reason for the lower protective effect in the formerly mentioned cases, but also the low levels of aberrations. The existence of a threshold for the activation of protective systems is quite possible, and the presented higher results which occur without affecting xeroderma pigmentosum cell cultures partially confirm the existence of such concentration-dependent, or clastogenic effect-dependent thresholds. This distortion displayed by the protec-

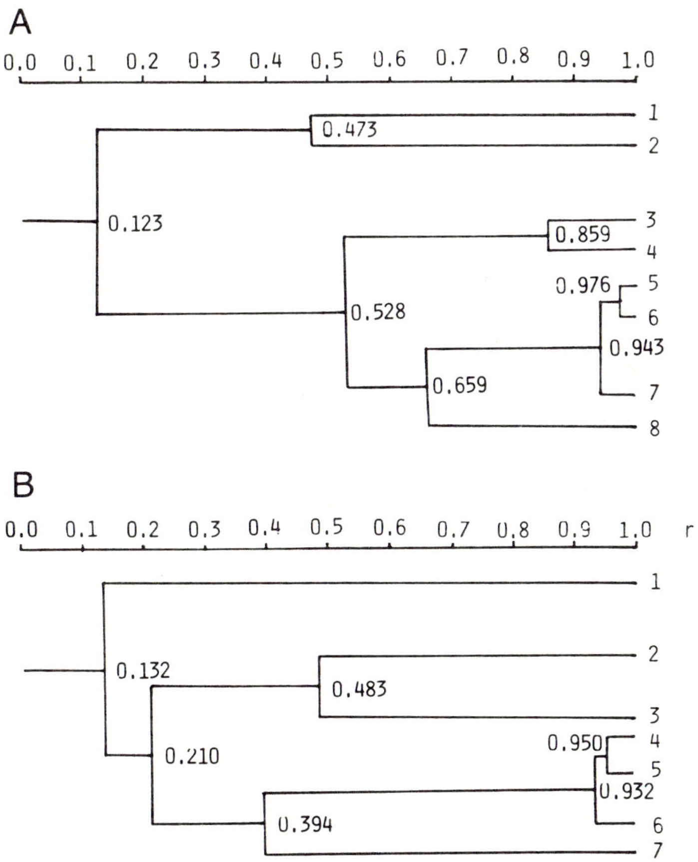

Fig. 8A,B. Dendrograms obtained on the basis of correlation matrices from: **A** 144 experimental variants (dipin and fotrin with and without anticlastogens WR 2721 and cystaphos). *r* Coefficient of correlation: *1* type of clastogen; *2* concentration of clastogen; *3* type of anticlastogen; *4* concentration of anticlastogen; *5* average quantity of single breaks per 100 cells; *6* average quantity of total breaks per 100 cells; *7* percentage of cells with chromosomal aberrations; *8* average quantity of double breaks per 100 cells. **B** 104 experimental variants (only combined treatment with mutagens and anticlastogens). *1* Type of anticlastogen; *2* type of clastogen; *3* concentration of clastogen; *4* average quantity of total breaks per 100 cells; *5* percentage of cells with chromosomal aberrations; *6* average quantity of single breaks per 100 cells; *7* average quantity of double breaks per 100 cells (Arutyunyan 1985)

tive mechanisms of the cell and induced by high levels of mutagens, in turn, may be one reason for the decrease in the protective effect when high levels of clastogenic action are observed.

Some theoretical explanations for the noted effects of protectors on the levels of chromosome aberrations, e.g., the activation of some of the endogenous, protective cell processes by the protector, can be considered. Such a possibility was analyzed above. Nonspecific mechanisms of protectors are described by the

Table 13. Comparative analysis of the number of damaged chromosomes in cells at different concentrations of thio-TEPA and addition of the anticlastogen WR 2721 (10^{-4} M) at the 28th h of lymphocyte cultivation (Arutyunyan 1985)

Treatment	Conc. of thio-TEPA ($\times 10^{-4}$ M)	Analyzed cells	Class			M	Distributions					
							Poisson			Pascal		
			0	1	2 and more		X^2	n	*p*	X^2	n	*p*
Thio-TEPA	0.25	200	193	3	4	0.06	1.97	1	>0.05	0.92	1	>0.05
1 h	0.375	200	188	11	1	0.065	0.03	1	>0.05	0.16	1	>0.05
	0.5	200	187	10	3	0.095	1.59	1	>0.05	0.40	1	>0.05
–	0.625	200	180	20	0	0.1	0.05	1	>0.05	1.01	1	>0.05
–	0.75	250	230	19	1	0.084	0	1	>0.05	0.52	1	>0.05
	1.0	100	79	20	1	0.22	6.18	2	<0.05	1.49	1	>0.05
	1.25	100	80	14	6	0.26	0.47	1	>0.05	0.06	1	>0.05
+ WR 2721	0.25	200	193	6	1	0.04	0.09	1	>0.05	0.01	1	>0.05
	0.375	200	193	6	1	0.04	0,09	1	>0.05	0.01	1	>0.05
	0.5	200	192	7	1	0.045	0.08	1	>0.05	0.03	1	>0.05
	0.625	100	94	6	0	0.06	0.01	1	>0.05	0.22	1	>0.05
	0.75	300	274	23	3	0.097	0.12	1	>0.05	0.28	1	>0.05

theories of a nonspecific reaction of cells to exogenous forces, developed by Eidus (1977), and by the theory of nonspecific protection by Spitkovskii et al. (1969).

These theories help to explain some of our data. It seems very reasonable that in every case and for every variant of the action of protectors there is a multidimensional, but nonspecific cell "stress"' which can not be reasonably explained by repair, the presence of endogenous thiols, or any other concrete mechanism.

What is a common characteristic in the coincidence of sensitivity of cell cultures to the S-dependent agents and anticlastogens? It is difficult to give one simple explanation. When the mechanisms of clastogenesis become more precisely known, it will become easier to understand the anticlastogenic effects and especially the time-effect mechanisms. However, the presented results, obtained with different classes of modifiers, which can either potentiate or decrease the action of clastogens by their addition in the same periods of the cell cycle, and on the basis of the presented test systems and analytic approaches, open new possibilities of comparing the effect of modifiers with better known molecular mechanisms on chromosomes.

3.2.2.3 Influence of Metabolic Activation

The in vitro systems considered above lack the important mechanisms of drug metabolism, e.g., activity in the liver of mammals. The development of effective and compatible protectors of the genetic material against the detrimental effects of mutagenic influences demanded testing under metabolizing conditions. Furthermore, practical in vivo systems were not always available for all studies on anticlastogenic action. Therefore, some research groups attempted to introduce the use of metabolizing fractions of mammalian liver (S9-mix) in their in vitro experiments. The applicability and reliability of this test system was documented by a series of studies in genetic toxicology testing (see Gebhart et al. 1984 for refs.). This latter group performed a series of experiments regarding the influence of S9-mix metabolism on anticlastogenic action in human lymphocyte cultures of β-aminoethylisothiouronium (AET), sodium fluoride (NaF) (Gebhart et al. 1984), and vitamins C and E (Gebhart et al. 1985). The model clastogens employed by these authors were trenimon (TR), cyclophosphamide (CP), and bleomycin (BLM). The clastogen, anticlastogen, and S9-mix were added simultaneously to the cultures for 2 h, 25 h before harvesting (in the BLM experiments 3 h treatment, 4 h before harvesting). The data of these experiments are summarized in Fig. 9.

In the absence of S9-mix, a distinct dose-dependent anticlastogenic action against trenimon was found for all the presumed anticlastogens, which, however, was clearly modified under the influence of rat liver S9-mix. Under these conditions the anticlastogenic potential seemed to be reduced in all experiments with TR and was nearly deleted in the experiments with activated cyclophosphamide, particularly in the case of vitamin E. BLM lost some of its strong chromosome-damaging effect under S9-mix activation. While AET acted as a coclastogen with BLM in cultures without S9-mix, its effect turned to anticlastogenic activity under activating conditions. The influence of NaF on the BLM-induced aberration rate

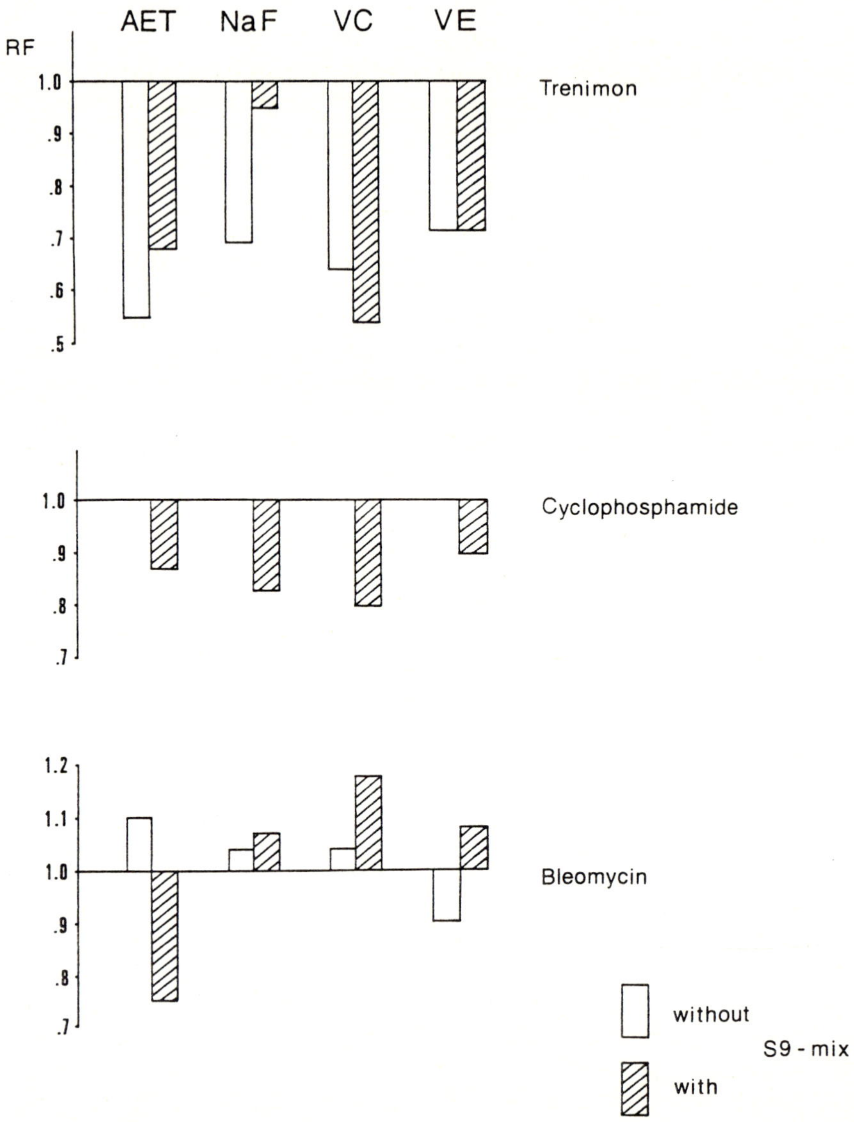

Fig. 9. Collected data from anticlastogen experiments with addition of S9-mix on human lymphocyte cultures. *AET* See above; *NaF* sodium fluoride; *VC* vitamin C; *VE* vitamin E. (Gebhart et al. 1984, 1985; see text for details)

remained unchanged, however, regardless of whether or not S9-mix was present. Neither vitamin significantly affected the chromosome-breaking activity of BLM if the S9-mix was absent, but they increased the clastogenicity of this cytostatic antibiotic under metabolizing conditions. The mechanisms responsible for the reduction of the anticlastogenic potential of AET and NaF under the influence of S9-mix were described by the authors as follows: (1) an increase in the clastogenic action by metabolic activation, as was evident in most TR experiments, without changing the absolute efficacy of the anticlastogens; (2) a deactivation of the protectors by the S9-mix; (3) a combination of these mechanisms; and (4) possible comutagenic action of S9-mix components, as observed by others (Saccone and Pariza 1981; Rossman and Molina 1983). Data presented above on the increase in chromosome-damaging activity of TR by the addition of S9-mix may be interpreted as favoring mechanisms 1 and/or 4. If this increase in clastogenicity is accompanied by a change in the spectrum of induced primary lesions, and if the anticlastogens are active on certain types of primary lesions only, as suggested by previous observations (Gebhart 1974, 1978; Gebhart and Becher 1976), this could also contribute to the effects observed in the reported study. This view may also be supported by the data on a decrease in the BLM-induced frequency of chromosome damage, coinciding with a change in AET activity from co- to anticlastogenic. The reason for the deactivation of BLM by metabolizing systems has recently been discussed by Trush et al. (1982), who after addition of lung microsomes at the molecular level also observed a decrease in the DNA damaging activity of BLM.

From a more practical point of view, the presented data on the reduction of anticlastogenic activity by metabolizing microsome preparations could lead to the conclusion that these findings might rule out any practical application of antimutagens in vivo. It should be taken into consideration, however, that these in vitro systems can only simulate in vivo activation in a very approximate way, and, furthermore, that long-term treatment is not possible is this model system. On the other hand, some very encouraging data on antimutagenic action from in vivo studies have been reported. AET displayed a distinct anticlastogenic action on γ-irradiated mice (Chaudhuri and Langendorff 1968; Garriott and Crowe 1983) as did S-2-(3-aminopropylamine)-ethyl-thiophosphoric acid on X-irradiated animals (Heddle and Harris 1975). Renner (1984) showed an in vivo anticlastogenic effect of the antioxidant, ethoxyquine, on the bone marrow of different species (see Sect. 3.3).

The toxic effects of known clastogens were reported to be distinctly diminished by AET (Asano et al. 1961) and N-acetylcysteine (Kline et al. 1973; Gurtoo et al. 1983; Harrison et al. 1983). Also, the therapeutic Reducdyn, which was previously shown to be anticlastogenic in human lymphocyte cultures (Stosiek and Gebhart 1974), has been employed as a radiation protector and a liver therapeutic for a long time.

These few examples document well that substances reported to be anticlastogenic from model in vitro systems also retain their biological activities under in vivo conditions. On the other hand, they throw some doubt on the reliability of the effects of S9-mix activation in human lymphocyte cultures. Further studies on this specific problem, however, may contribute to the development of standardized in vitro systems for the examination of antimutagenic action on a more reliable and practical basis, which for instance, should also include the findings presented in Section 3.2.2.1.

A distinct dose-dependent protective effect of ascorbic acid was detected in the clastogenicity of long-term treatment (24 h) with trenimon if the vitamin was added to the cultures simultaneously or before the clastogen (Gebhart et al. 1985). In short-term tests (2 or 3 h) of clastogen exposure ending 23 and 21 h before harvesting), simultaneous addition of vitamin C or E reduced the clastogenic action of trenimon, regardless of whether S9-mix was present or absent. While vitamin C also decreased the frequency of chromosome damage induced by S9-mix-activated cyclophosphamide, vitamin E was inactive under the same conditions. Neither vitamin significantly affected the clastogenic activity of bleomycin in the absence of metabolizing S9-mix, but they increased it under metabolizing conditions. This coclastogenic action, according to the authors' assumptions, cannot be attributed to their antioxidant activity but rather to the changed action of BLM (Gebhart et al. 1984), or to the differing activity of both vitamins under these conditions, as was suggested by the outcome of experiments with alkylating agents. Although these speculations must be investigated further, the authors summarized their findings: "The clear anticlastogenic effect of both vitamins on the action of alkylating agents ... additionally supports the idea of the possibility of practical application of natural protectors against the clastogenic (mutagenic/carcinogenic) action of chemical mutagens. Although metabolizing in vitro conditions apparently can alter this activity, the in vivo data presented so far on antimutagenic ..., anticarcinogenic, ..., and antitoxic action of vitamins can be interpreted in favour of this practical application."

3.2.2.4 Interphase Studies Using the PCC Technique

The usual techniques pertaining to metaphase analysis of induced chromosome damage and its reduction by anticlastogens allow only rather indirect conclusions on the primary events taking place in the previous interphase. It is well known, however, that various repair systems act on these primary events and, in addition, selection can blur the original picture during the time which elapses between the application of a test agent and the emergence of scorable metaphases. Therefore, attempts to obtain more sophisticated insights into the actual primary processes of anticlastogenic action should be aimed at interphase events and should not be based on metaphase analysis only.

A few investigators have tried successfully to overcome the shortcomings of "indirect" metaphase analysis by using the technique of "premature chromosome condensation" (PCC) of mutagen-exposed cells (Hittelman and Rao 1974, 1975; Sognier et al. 1979; Hittelman and Pollard 1982). This phenomenon is induced by the fusion of the interphase target cells with mitotic "inducer" cells, as primarily reported by Johnson and Rao (1970).

Despite evident advantages, only few data from PCC studies on the induction of chromosomal changes during interphase have been reported thus far.

For instance, the clastogenic activities of alkylating agents (Hittelman and Rao 1974; Hittelman and Pollard 1982; Johnson et al. 1982; Abkaí et al. 1987), but also of the cytostatic antibiotic bleomycin (Sognier et al. 1979; Abkaí et al. 1987) were studied on interphase stages. The main result was that a large amount of the induced chromosomal damage detectable at various interphase cells disappeared until the exposed cells reached mitosis. The striking differences between the aberration yields in metaphase and PCC (see Fig. 11A) have been explained by repair processes which, apparently, are active in G_2 phase (Sognier et al. 1979; Abkaí et al. 1987), but also in G_1- or S-phase (Clarkson and Humphrey 1976; Sen and Hittelman 1984). In addition, these findings support the assumption that selective factors also act on heavily damaged cells, thus preventing them from entering mitosis (Tobey 1975; Rao and Rao 1976).

With these facts in mind, Tittelbach et al. (1989) attempted to analyze anticlastogenic action in PCC. As the chromosome number of the target cells could be one of the technical limitations of this method (Sognier et al. 1979; Hittelman and Pollard 1982; Pantelias and Maillie 1984; Abkaí et al. 1987), the authors chose a fibroblast strain of the male Indian muntjak which has only seven well-discernible chromosomes (Fig. 10).

The comparative analysis of interphase and metaphase data of their study revealed striking changes in the trends of the co- and anticlastogenic action of the sulfhydryl compounds N-acetylcysteine (NAC) and β-aminoethylisothiouronium (AET). While interphase data from simultaneous short-term administration of bleomycin and NAC revealed a coclastogenic effect of the latter, no striking effect at all was seen in the metaphases. An additional, consecutive application of NAC for 1 h deleted the coclastogenic effect and even resulted in an anticlastogenic response. Previous metaphase studies on proliferating lymphocytes in vitro revealed a coclastogenic action of NAC when administered simultaneously with bleomycin, and a protective effect was seen when a posttreatment protocol was followed (Kartzewski and Gebhart 1987).

Using AET, Tittelbach et al. (1989) found no clear-cut effect on G_2-PCC, but a coclastogenic action on G_1-PCC when it was applied for 1 h after bleomycin treatment. These data did not correspond to those obtained with NAC, although both compounds were supposed to act with similar mechanisms. Previous studies on metaphases of human lymphocytes which were subjected to a 3-h treatment with bleomycin and AET also resulted in a coclastogenic action of the latter (Gebhart et al. 1984). In contrast to their data obtained from cells which presumably were mainly in G_1 at the time of exposure, the metaphase data of Tittelbach et al. (1989) obtained from muntjak fibroblasts did not show any difference regardless of whether or not they were derived from G_1- or G_2-exposure. As a repair-inhibiting agent, the cycloheximide derivative streptovitacin A (SA) was studied by the same authors to examine its effect on bleomycin-induced chromosome damage in muntjak fibroblast interphase. A striking decrease in the chromosome breakage rate in G_2-PCC and in metaphases was noted after treatment in G_2-phase (Fig. 11B). A coclastogenic action of SA and bleomycin, if administered in G_1, thus far, has not been substantiated by a sufficient number of experiments. The authors concluded from their results that the use of the PCC technique for anticlastogen studies, although a very complex assay, could provide a more sensitive method than conventional metaphase analysis. Far from becoming a routine technique, however, it could offer a variety of possibilities to elucidate immediate clastogenic effects and their modification by interacting compounds. Certainly a large series of further PCC experiments must be performed before a definite judgement of the value of this method for a broader use in anticlastogen research becomes possible.

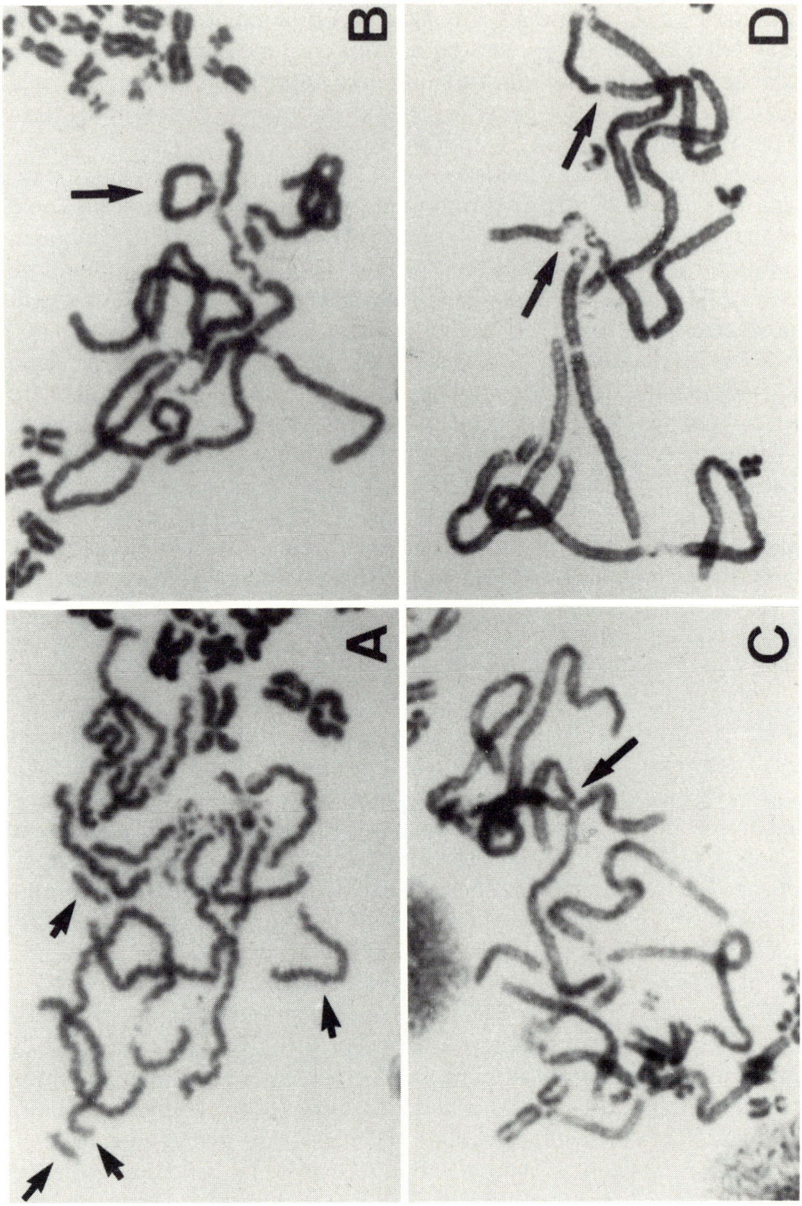

Fig. 10. Chromosome aberrations (*arrows*) in prematurely condensed G1-phase (**A**) and G2-phase (**B-D**) chromatin (PCC) (Photograph: Tittlebach). **A** Various fragments; **B** centric ring; **C** chromatid inte rchange; **D** chromatid breaks

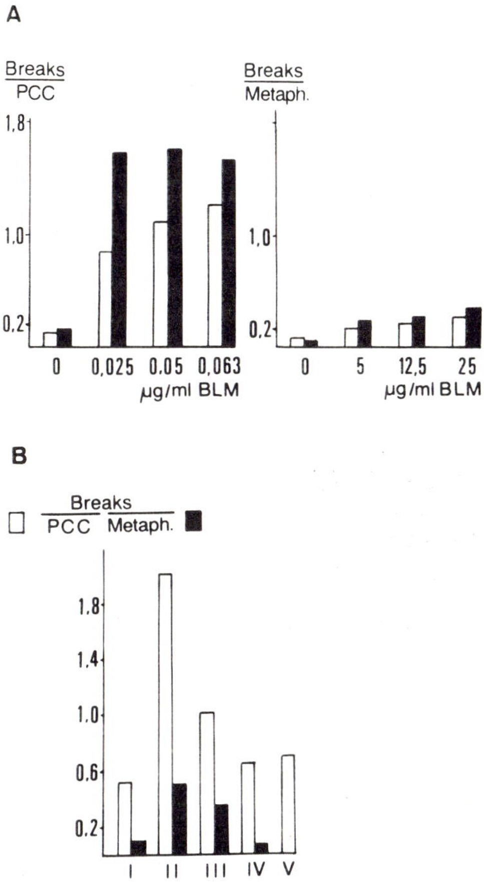

Fig. 11A,B. Data from PCC analyses on clastogenic and anticlastogenic action Tittelbach et al. 1989). **A** Differences in aberration yields between metaphases and PCC. Indian muntjak fibroblasts were exposed to bleomycin (*B LM*) for 30 min and immediately followed by cell fusion for the analysis of PCC. *White columns* data from G1- PCC; *black columns* data from G2-PCC. Metaphase analysis was performed on cells treated for 30 min with BLM in G1- (*white columns*) or G2- (*black columns*) phase. **B** Anticlastogenic action of streptovitacin A in G2-PCC and metaphase. *I* Untreated controls; *II* BLM 0.025 µg/ml for 30 min followed immediately by cell fusion; *III* BLM 30 min, cell fusion 60 min after the exposure was completed; *IV* BLM + streptovitacin A (SA: 25 µg/ml, cell fusion after further 60 min of SA exposure); *V* 90 min SA exposure followed by cell fusion

3.3 Anticlastogenic Action in Vivo

3.3.1 Studies on Mammals

A series of mammalian in vivo test systems has become an indispensable tool in modern genetic toxicology because of their metabolic proneness to the human organism, their ability to detect a variety of genetic damage, and their high reliability. It was not, however, until recently that these systems were also involved extensively in anticlastogen research.

Previous studies (Šrám and Kocisová 1975) showed that rifampicin and chloramphenicol diminished the ability of TEPA to induce X- or Y-univalents in spermatocytes of mice if applied on spermatogonia. Nemirovsky and Klimenko (1973) reported on an anticlastogenic effect of posttreatment with caffeine to liver cells of rats which were previously injected with the alkylating agent dipin. The effect of polyvinylpyrrolidone (PVP) on the mutagenic activity of the antitumor agent photrin (= 2,2,4,4,6-pentaethylene-imino-6-morpholinocyclophosphazatriene) was determined by Revazona and Radchenko (1976) using the "dominant lethal test" in mouse, which eventually also detects cytogenetic damage. At therapeutic doses PVP exerted an antimutagenic effect, as the percentage of postimplantation deaths and of dead embryos was lower in animals given photrin plus PVP than in those provided with photrin alone.

Using the micronucleus assay, Zimmering et al. (1975) found a clearly dose-dependent reduction of the number of TEM- or hycynthone-induced micronuclei after pre- and posttreatment with a pyrogen.

More recently this assay, which is regarded as suitable for detecting the consequences of clastogenic activity, was also used for a number of further anticlastogen studies. In addition, with these tests several natural substances were examined with respect to their protective potential versus cytogenetic damage produced by a variety of mutagens (Table 14).

A clear protective effect against the clastogen-induced formation of micronuclei was reported from bioantimutagens like l-cysteine, ascorbic acid, β-carotene, α-tocopherol, vanillin, resorcinol, and pyrogallol, but also for juices or extracts from plant food like spinach and carrots, or even betel leaves.

In most of these studies a more or less simultaneous application of clastogen and protector had been chosen. No detailed attention was given to the metabolic mechanisms or mechanisms of action within the test organisms, but the outcome of a reduced number of micronuclei seemed sufficient to the authors to define an anticlastogenic effect of the respective bioantimutagens.

A more analytic method, however, was chosen by some other groups who, on the one hand, compared various endpoints of clastogenic action, or different test systems and test clastogens. On the other hand, they tried to determine whether the anticlastogenic effect is dependent on the chemical structure of the anticlastogens used, or they tested several ways of treatment (pre-, post-, and simultaneous treatment).

Of particular interest are the extensive studies of Renner and his group, who, on the one hand, compared the efficacy of certain anticlastogens on different

Table 14. In vivo studies on anticlastogenic action in mammals

Anticlastogen	Clastogen(s)	Test system[a]	Effect (%)	Reference[b]
l-Cysteine	Diiodohydroxyquinoline	MBM/Mn	>50	1
Glutathione	DMBA	MBM/Mn	50	3a
"	Benzo<α>pyrene	MBM/Mn	28	3a
"	Cyclophosphamide	MBM/Mn	51	3a
"	Mitomycin C	MBM/Mn	27–74	3a
Ascorbic acid	Diiodohydroxyquinoline	MBM/Mn	+	2
β-Carotene	Benzo<α>pyrene,	MBM/Mn	41–61	3b
"	Mitomycin C		44–71	3b
"	Thio-TEPA	CBM/Ch	>50	4
"	Methylmethanesulfonate	CBM/Ch	>80	4
"	Busulfan	CBM/Ch	>80	
α-Tocopherol	DMBA	MBM/Mn	50–70	5
"	Methylmercury	CF/Ch	>80	6
"	Cyclophosphamide	MBM/Ch	–	7
"	Benzo<α>pyrene	MBM/Ch	+	28
Vanillin	Mitomycin C	MBM/Mn	50	8
Resorcinol + Pyrogallol	Benzo<α>pyrene	MBM/Mn	+	9
Indomethacin	Benzene	MBM/Mn	+	10
7,8-Benzoflavone	DMBA	MBM/Mn	90	11
"	Benzo<α>pyrene	MBM/Mn	75	11
Na-selenite	DMBA	MBM/Mn	54	12
"	Benzo<α>pyrene	MBM/Mn	30	12
"	Mitomycin C	MBM/Mn	28–78	12
Pyridine	Benzene	MBM/Mn	up to 80	13
"	Benzo<α>pyrene	MBM/Mn	>50	13
"	Cyclophosphamide	MBM/Mn	–	13
Propylgallate	DMBA	MBM/Mn	50–70	5
β-Sitosterol	DMBA	MBM/Mn	50–70	5
Ethoxyquine	Cyclophosphamide	CBM/Mn	>50	14a
	"	RBM/Mn	>50	14a
	"	MBM/Mn	>50	14a
	"	CBM/Ch	>80	14a
	"	MDL	+	14b
	"	MSG/Ch	+	14b
Methyl esters of:				
Tauric acid	Busulfan	CBM/Ch	>60	15
Tridecanoic acid	"	"	"	15
Myristic acid	"	"	"	15
Pentadecanoic acid	"	"	"	15
Palmitic acid	"	"	"	15
Maragaric acid	"	"	"	15
Stearic acid	"	"	"	15
Oleic acid	"	"	"	15
Linoleic acid	"	"	"	15
α-Linolenic acid	" , Thiotepa	"	" ,85	15,31
n-Nonodecanoic acid	"	"	"	15
Gamma-linolenic acid	"	"	–	15
Arachidonic acid	"	"	–	15
Mercaptopropionylglycine (Thiola)	Benzo<α>pyrene	MBM/Ch	>50	16
D-penicillamine	Cyclophosphamide	MBM/Ch	20–100	17

Table 14 *(continued)*

Anticlastogen	Plastogen (s)	Test system[a]	Effect (%)	Reference[b]
Ethanol	Patulin	CBM/Ch	>60	18
	Aflatoxin B, cyclophosphamide	CBM/Ch	–	18
Naphthoflavones	DMBA	MSG/Spa	50–80	5
Polyvinylpyrrolidone	Fotrin	MDL	+	19
Dimethylsulfoxide	Benzene	MBM/Mn	>80	33
Toluene	Benzene	MBM/Mn,Ch	50–75	23
"Carrot" extract	Cyclophosphamide	MBM/Mn	50	20
"Spinach" extract	Cyclophosphamide	MBM/Mn	40	20
Vegetable juices	DMBA	RBM/Ch	28	21
Betel leaf extr.	Nitrosamines	MBM/Mn	+	22
Hydroxychavicol	Tobacco-specific	MBM/Mn	+	29
"	Nitrosamines	"	+	29
Caffeine	Dipine	RLC/Ch	+	24
Rifampicin,	TEPA	MSP/uv	+	25
chloramphenicol	"	"	+	25
"Pyrogen"	TEM, hycanthone	MBM/Mn	+	26
Phenobarbital,	DMBA	MDL	>80	27
methylcholanthrene	"	"	>80	27
Coumarin	Benzo<α>pyrene	MBM/Mn	52	30
Tannic acid	Mitomycin C, ethylnitrosourea NQO	MBM/Mn	>60	32

[a]Abbreviations: MBM, mouse bone marrow; CBM, Chinese hamster bone marrow; RBM, rat bone marrow; MDL, mouse dominant lethal test; MSG, mouse spermatogonia; RLC, rat liver cells; MSP, mouse spermatocytes; Mn, micronucleus test; Ch, chromosomal aberration test; uv, univalents; Spa, sperm anomalies.

[b]References: *1*, Ghaskadbi et al. (1987); *2*, Ghaskadbi and Vaidya (1989); *3a*, Raj and Katz (1985a); *3b*, Raj and Katz (1985b); *4*, Renner (1985); *5*, Raj and Katz (1984a,b); *6*, Gilbert et al. (1983); *7*, Kodytkova et al. (1980); *8*, Inouye et al. (1988); *9*, Paschin et al. (1986); *10*, Pirozzi et al. (1989); *11*, Raj and Katz (1983); *12*, Raj and Katz (1986); *13*, Harper and Legator (1987); *14a*, Renner (1984); *14b*, Renner and Knoll (1984); *15*, Renner (1986); *16*, Galdean et al. (1986); *17*, Nair and Vogel (1989); *18*, Korte et al. (1979); *19*, Revazova and Radchenko (1976); *20*, Abraham et al. (1986); *21*, Ito et al. (1986); *22*, Padma et al. (1989); *24*, Gad-El-Karim et al. (1984); *23*, Nemirovsky and Klimenko (1973); *25*, Srám and Kocisová (1975); *26*, Zimmering et al. (1975); *27*, Murota and Shibuya (1983); *28*, Rao et al. (1986); *29*, Amonkar et al. (1989); *30*, Morris et al. (1986); *31*, Renner (1990a); *32*, Sasaki et al. (1990b); *33*, Au et al. (1990).

endpoints and various test systems, and, on the other hand, in a large series of tests, could relate the anticlastogenic potency of a series of methyl esters of fatty acids to their chain length. Using the chromosome aberration test on Chinese hamster bone marrow, he found that the clastogenic activity of the alkylating cytostatic busulphan was not modulated by the short-chain fatty acids, but that the fatty acids from lauric acids (C12) up to nonodecanoic acid (C19) reduced the rate of aberrant metaphases from 9.4 to about 3% at doses of 100 mg/kg and less. Other chemical properties of the fatty acids (saturated or unsaturated, number of double bonds, even- or odd-numbered) had no influence on the anticlastogenic effects. The only exceptions to this rule were arachidonic acid, which had no effects, and α linolenic acid, which had no consistent effect on the action of busulphan (Renner 1986).

The anticlastogenic effect of ethoxyquine was tested comparatively on bone-marrow cells of Chinese hamsters using three different endpoints (chromosome aberrations, micronuclei, and SCEs). The micronucleus system was used to compare this effect on three different species (Chinese hamster, mouse, and rat). While ethoxyquine did not influence the induction of SCE by cyclophosphamide, it drastically reduced the micronucleus rate induced by this cytostatic at all doses tested. This action was also found in three different mouse strains (NMRI, C3H, C57B6) and in the rat. In the chromosome aberration test a distinct anticlastogenic effect of ethoxyquine was detected which was similar to that found on micronuclei. At the highest doses of ethoxyquine (100 mg/kg) the cyclophosphamide-induced damage was even reduced to the level of the spontaneous rates (Renner 1984). These data were substantiated by respective studies on chromosome aberrations in spermatogonia of mice as well as by the dominant lethal test (Renner and Knoll 1984). Reviewing all his data, Renner (1990b) concluded that the content of the diet of specific fatty acids may be associated with well-known epidemiological facts, e.g., the differences in cancer rates between various populations, but also the differences in the frequencies of cardiovascular diseases.

Most recently, using the mouse micronucleus assay in combination with the biochemical detection of metabolites, Au et al. (1990) found a diminishing effect of dimethyl sulfoxide on benzene-induced endpoints, which was accompanied by changes in the amount of the metabolites phenol and trans-transmuconic acid.

As shown in Table 14, a series of studies using in vivo systems other than micronucleus assays, in particular chromosome studies on the bone marrow of exposed animals, also detected anticlastogenic action of natural antimutagens, e.g., β-carotene, α-tocopherol, retinol, fatty acids, vegetable juices, and naphtoflavones, but also of synthetic agents like PVP, penicillamine, mercapto-propionylglycine, and ethoxyquine. There is, however, some evidence that the induction of liver enzymes can distinctly influence the clastogenic activity of mutagens/carcinogens. Cytochrome P450 isozyme was suspected to be one of these responsible enzymes (Harper and Legator 1987) on the basis of comparative studies on the influence of pyridine and methylchonlathrene on micronucleus induction by benzene, benzopyrene, and cyclophosphamide. The same enzyme had been suggested to be responsible for the modifying effects of the enzyme inducers phenobarbital and 3-methylcholanthrene on dominant lethal events induced by 7,12-dimethylbenz<α>anthracene in mice (Murota and Shibuya 1983).

Furthermore, spontaneous chromosomal instability can be reduced to the normal value in vivo by repeated injections of bovine superoxide dismutase, as shown in New Zealand black mice (Emerit et al. 1981). The decrease in aberration rates occurred in all treated animals and the enzyme appeared to be effective for all types of aberrations.

Despite their highly estimated reliability in genetic toxicology, mammalian in vivo test systems have not found popular use in anticlastogen studies. This, in part, may be due to the high complexity of these systems per se, but also to the immense efforts these studies demand. However, future antimutagen research certainly cannot do without these systems, and therefore, extended efforts must be invested into such studies than has been done so far.

3.3.2 Studies on Humans

The research on anticlastogens effects in man has two main directions. One is the protection of normal cells during cancer chemo- or radiotherapy, radioprotection being the most developed field (see Phillips and Wasserman 1984). The second, thus far rather underdeveloped direction, is chemoprevention of cancer or mutations in high risk groups with a parallel decrease in clastogenic effects.

Emerit et al. (1979) reported that the treatment with the drug D-penicillamine (dimethylcysteine) reduced the incidence of chromosome breakage in the lymphocyte cultures of three patients with Crohn's disease with breakage frequencies of 37.8, 30.3, and 23.7%, respectively, to normal values after 2 months of treatment. One of the authors' hypotheses as to the nature of the effect is the coordination of the drug with copper ions, i.e., the formation of copper chelate, which can act as a superoxide dismutation agent.

In a series of studies by Šrám et al. (1983a,b, 1985, 1986) it was demonstrated that the systematic administration of ascorbic acid in various groups of individuals with genetic risks can decrease the levels of their aberrations. This was shown for workers exposed to halogenated ethers (Šrám et al. 1983a), coal-tar workers (Šrám et al. 1983b), and soft coal mining (Šrám et al. 1985), or workers exposed to gaseous substances (Šrám et al. 1986).

In the work of Šrám et al. (1983a) the decrease in chromosomal aberrations was demonstrated to be due to the chromosome-type breaks which were decreased twofold after treatment with ascorbic acid. The possible mechanisms of its protective effect were discussed earlier (see Sect. 3.1). Rössner et al. (1988) investigated the personnel of a chemical laboratory exposed to cytostatic drugs, who were supplemented with ascorbic acid for 6 months. Ascorbic acid did not significantly decrease the levels of their aberrant lymphocytes. The authors noted that their result, in contrast to that obtained by Šrám et al. (see above), "seems to imply that success of ascorbic acid prophylaxis is apparently dependent on the class of mutagen involved." They reported that their investigated risk group was mainly exposed to directly acting mutagens, while the action of ascorbic acid may be based on a metabolic biotransformation process. In the same paper these authors presented interesting results obtained by the so-called nucleolus test, which — according to the authors -- can indicate an inhibition of rRNA synthesis (Smetaná et al. 1969, 1980). After ascorbic acid treatment, the number of inactivated nucleoli, which was higher in the exposed group (33.4%) than in the controls (24.3%), decreased to 20.5%. Based on this fact, the authors recommended the application of the nucleolus test for the estimation of environmental mutagenic and carcinogenic action. The data on anticlastogenesis in man are not only limited to the study of chromosomal aberrations. In a few groups having increased genetic and carcinogenic risk, a significant reduction in the levels of micronuclei in oral mucosa cells of betel quid/tobacco chewers (Stich et al. 1984) and snuff dippers (Stich and Rosin 1985) was shown by the oral administration of vitamin A or β-carotene. Recurrence of the prior micronuclei levels, however, was observed if the vitamin administration was ceased (Stich et al. 1985). The specific methodological approach for "analyses of micronuclei in buccal smears" yields good and reliable estimates of clastogenic activity and its modification in environmental mutagen monitoring (Stich 1986). Its suitability for studying anticlastogenic/anticarcinogenic effects was recently summarized and critically evaluated by Rosin (1990).

The number of experiments using human lymphocyte cultures to study the modification of cytogenetic changes, which were induced in these cells in vivo, is increasing steadily. For example, Nordenson (1977), by adding the enzymes superoxide dismutase and catalase to lymphocyte cultures, succeeded in decreasing the levels of chromatid breaks and gaps, and in an experimental variant, taking into account the combined action of the enzymes, also decreased the levels of chromosome aberrations in the lymphocyte cultures from two patients with Fanconi's anemia. Nordenson (1977) noted that he had also shown the analogous protective effect of cysteine in the lymphocyte cultures obtained from patients with Fanconi's anemia. As mentioned earlier (see Sect. 3.2.1.2.2), a protective effect of l-cysteine was also shown for the lymphocyte cultures from scleroderma patients by Emerit et al. (1974). A significant protective effect of D-penicillamine was revealed after its addition to the blood cultures of patients with Crohn's disease (Emerit et al. 1979). The protection was significant for breaks, fragments, and gaps. It is important to note in this context that protective effects were demonstrated for the different levels of aberrations in lymphocytes cultures that were set up on modified TCM 199 and RPMI 1629, distinctly differing in their l-cysteine content (0.1 and 30 mg/l, respectively).

A "vice-versa" assay was used by Pohl and Reidy (1989) who studied the sensitivity to the chromosome-damaging effect of bleomycin of lymphocytes obtained from individuals previously supplemented with 1 g vitamin C per day. This supplementation decreased the amount of bleomycin-induced chromosome damage. The authors did not observe such changes in lymphocytes from control individuals (no vitamin C) sampled at the same time. From these findings the authors concluded that it might be prudent to consider dietary and perhaps other life-style factors when interpreting the results from this bleomycin assay and related assays regarding genetic instability.

For a better understanding of the in vivo mechanisms of protector effects in man, it is reasonable to compare their cytogenetic effects with those found in experimental test systems. However, papers reporting comparisons of this kind are rare. They are more numerous in the research on clastogenesis. From our point of view, the investigations of Norppa et al. (1980a,b,c) on the action of sodium selenite on the aberrations and SCE in human lymphocytes (Norppa et al. 1980a) are of particular interest. This is also valid for aberrations in primary bone marrow cells, primary spermatocytes (Norppa et al. 1980b), and aberrations and SCE in Chinese hamster bone marrow (Norppa et al. 1980c).

The following results were obtained. The administration of sodium selenite in groups of patients and healthy persons produced no effect regarding the increase in aberrations or SCE. Injections yielded no effect on the level of chromosomal aberrations in mouse bone marrow or primary spermatocytes. However, in Chinese hamster bone marrow very high doses of selenium induced pronounced aberrations and SCE. Thus, the authors suggest that "as selenium, however, seems to be an essential trace element (Schwarz 1976), a mechanism may exist in vivo which prevents the harmful effects of Se at low levels."

It is difficult to imagine that the action of environmental agents depends on the natural mechanisms of protection versus specific mutagens at low levels, but this concept presents a productive approach to the search for reserves of natural anticlastogenic mechanisms.

In summary, the scarce data obtained from the in vivo observations in man are insufficient to draw concrete conclusions on the necessity of applying anticlastogens in individuals with increased levels of cytogenetic changes. It is especially worthwhile, however, to continue research on the utility of anticlastogens in those groups in which cytogenetic changes are associated with serious diseases.

3.4 Action of Anticlastogens on Sister Chromatid Exchange (SCE)

The extension of anticlastogen studies to other relevant endpoints seemed desirable as a means for obtaining better insights into the mechanisms of anticlastogenic action. As extensively discussed in Section 2.3, in the 1970s, the sister chromatid exchange test gained considerable attention as an assay using a new endpoint of genotoxicity. At that time, indications increased that SCEs represented an endpoint of primary lesions in DNA differing from that represented by classical structural chromosome aberrations. As the first studies on the action of anticlastogens on SCE supported this view, the data collected in this field will be discussed in the following.

As shown in Table 15, several classical anticlastogens, such as sulfhydryl compounds, vitamins, and other bioantimutagens, were studied with respect to their ability to modify the SCE-inducing activity of mutagens. Several were shown to decrease the yield of SCE, particularly if the SCEs were induced by clastogens acting via the production of free radicals or monofunctional alkylation (Speit et al. 1980a,b; Sono and Sakaguchi 1981a,b).

In some cases an anticlastogenic effect was found for substances which did not reduce the SCE-inducing effect of chromosome-damaging agents (Speit et al. 1984). On the contrary, even an increase in the SCE frequency was found when mitomycin-C-pretreated CHO cells were cultured in the presence of antimutagenic flavorings (vanillin, ethylvanillin, anisaldehyde, cinnamaldehyde, coumarin, and umbelliferone: Sasaki et al. 1987). All these compounds have either an α,β-unsaturated carbonyl group or a carbonyl functionality adjacent to the phenyl group which, according to these authors, may react with an enzyme SH group and cause

Table 15. Action of anticlastogens on the chemically induced SCE frequency in mammalian and human cell systems

SCE-affecting agent	Inducing clastogen	Cell system[a]	Effect	Reference[b]
L-cysteine	Hydrazine, isoniazid	V79	+	1
	Ascorbic acid	V79	+	2
	Nitrosocimetidine	HLY	+	3
	X-ray	HLY	+	4
	Hydrogen peroxide	V79	–	5
	Trenimon, 8-HQS	HLY	–	6
	UV	HLY	–	5
Cysteamine	Nitrosocimetidine	HLY	+	3
	UV	CHO	+	7
AET	Trenimon, 8-HQS	HLY	–	6
	Cyclophosphamide/S9	HLY	–	8
Glutathione	Ascorbic acid	V79		2
	Hydrogen peroxide	V79	–	2
	Nitrosocimetidine	HLY	–	3
Homocysteinethiolactone	Trenimon, 8-HQS	HLY	-	6
Amino acids	Cyclophosphamide	RBM	+	9
Retinol (vitamin A)	Cyclophosphamide	V79	++	10
	Aflatoxin B1	V79	+	10,11
	DMN/DEN	V79	++/(+)	12
	Benzo<α>pyrene	V79	–	11
	Mitomycin C	V79	–	13
Synthet. retinoids	Mitomycin C	V79	–	13
Ascorbic acid (vitamin C)	MNNG	CHO	+	14
	Thio-TEPA, 1-ethionine	HLY	(+)	15
	Cyclophosphamide, Mitomycin C	MBM	+	16
	Nitrosocimetidine	HLY	–	3
	Trenimon, cyclophosphamide	HLY/S9	–	17
α-Tocopherol (vitamin E)	Nitrosocimetidine, Trenimon,	HLY	–	3
	HQS, cyclophosphamide	HLY/S9		17
Selenite	MMS, N-HAAF	HLY	+	18
	Methyl mercury, mercuric chloride	HLY	+	19
	MNNG, MNU	V79	+	20
Spermidine	Mitomycin C	CHO	+	21
	MNNG		–	21
Carrot juice	Cyclophosphamide	CHO	+	22
		HLY	(+)	22
Chlorophyllin	"Airborne pollutants"	HLY	++	23
Colchicin	EMS	DON6	+	24
Indomethacin	Benzidine	RH4	++	25
		HEPG2	++	25
	Benzo<α>pyrene, DMBA	AH66B	(+)	26
		R1	(+)	26
Antipain	MNNG	HLY	+	27
	3-Aminobenzamide	CHO	+	28
Amastatin, betastatin,, chymostatin, elastatinal	MNNG	HLY	+	28
	MNNG, ENNG	HLY	+	28
Leupeptin	MNNG	HLY	(+)	28

Table 15. *(continued)*

SCE-affecting agent	Inducing clastogen	Cell system[a]	Effect	Reference[b]
Aminoharman	Mitomycin C	OH-TRP2	+	29
	N-mustard-N-oxide	HNL3	++	29
	MNNG, NQO		+	29
	MMS		–	29
Norharman	Mitomycin C, MNNG	CHO	(+)	30
Catalase	Ascorbic acid	CHO/HLY	+	14
Cycloheximide	EMS	DON6	+	31
	UV	V79	++	32
Puromycin	EMS	DON6	+	33
Superoxide dismutase	TPA	CHO	+	34
Tannic acid	Mitomycin C	CHO	+	35
Trimethylhydroquinoline	Cyclophosphamide-con taining plasma from leukemic patients	HLY	+	36
Fluoride	Trenimon, 8-HQS, Cyclophosphamide	HLY/S9	–	8
TPA	Cyclophosphamide	V79	++	37
Vincristine	EMS	DON6	+	24
(2R,5R)-6-heptyne-2,5-diamine	Cis-platin	RBT9L	+	38
Krestin, lentinan	Mitomycin C	MBM	+	39

[a]Abbreviations: see previous tables and text; DMN, dimethylnitrosamine; DEN, diethylnitrosamine; MMS/EMS, methyl-/ethyl-methanesulfonate; N-HAAF, N-hydroxyacetylaminofluorene; MNU, methylnitrosourea; TPA, tetradecanoylphorbolacetate/myristate.

[b]References: *1*, Speit et al. (1980a); *2*, Speit et al. (1980b); *3*, Inoue et al. (1985); *4*, Abramovsky et al. (1978); *5*, Speit et al. (1982, 1984); *6*, Gebhart and Kappauf (1980); *7*, Bianchi et al. (1982b); *8*, Gebhart et al. (1984); *9*, Chen and Lin (1987); *10*, Huang et al. (1982); *11*, Quin et al. (1985); *12*, Huang (1987); *13*, Sirianni et al. (1981); *14*, Galloway and Painter (1979); *15*, Lialiaris et al. (1987); *16*, Krishna et al. (1986); *17*, Gebhart et al. (1985); *18*, Ray et al. (1978); *19*, Morimoto et al. (1982); *20*, Hu (1985); *21*, Cozzi et al. (1989); *22*, Darroudi et al. (1988); *23*, Hadnagy and Seemayer (1988); *24*, Sono and Sakaguchi (1982); *25*, Grady et al. (1986); *26*, Abe (1986); *27*, Umezawa et al. (1979, 1980); *28*, Schwartz and Weichselbaum (1985); *29*, Tohda et al. (1984); *30*, Rutten and Wilmer (1986); *31*, Sono and Sakaguchi (1981a); *32*, Ishii and Bender (1980); *33* Sono and Sakaguchi (1981b); *34* Nagasawa and Little (1981); *35*, Sasaki et al. (1989); *36* Raposa et al. (1982); *37*, Hsueh et al. (1981); *38* Milam et al. (1989); *39* Hasegawa et al. (1989).

higher-order structural changes. SCE-enhancing effects of vanillin were further observed using ethylmethanesulfonate, ethylnitronitrosoguanidine (ENNG), and ethyl- or methylnitrosourea as inducers of SCE. On the other hand, no influence of this flavoring on methylmethanesulfonate- or methylnitronitrosoguanidine-induced SCE rates could be detected. The authors, therefore, suggested that the SCE-enhancing effects of vanillin were dependent on the "quality" of lesions induced on DNA. These data, together, support the conclusion that the anticlastogenic effect in these cases is due rather to a specific prevention of break formation than to direct inactivation of the mutagen. Furthermore, this was regarded as an indication of the different mechanisms causing chromosome aberrations and SCE (Gebhart 1981). In such cases, SH compounds reduced the number of SCEs induced by oxidative agents; thus, a direct chemical interaction between protector and clastogen seems probable.

There are, however, also differences between anticlastogens concerning their ability to reduce the frequency of SCEs induced by the same clastogens, as shown, e.g., for UV light and nitrosocimetidine (Bianchi et al. 1982b; Speit et al. 1982; Inoue et al. 1985). Various responses to the modifying action of norharman were also reported to depend on the nature of the SCE-inducing mutagen. The number of SCEs induced in CHO cells by the directly acting mutagens, mitomycin C and MNNG, decreased in the presence of norharman, while a synergistic effect of this compound was found it combined with the indirect mutagens, 2-acetylaminofluorence, 2-aminofluorene, N-hydroxyacetylaminofluorene, and benzo<α>-pyrene (Rutten and Wilmer 1986).

On the other hand, sodium selenite was shown to reduce the number of SCEs induced by substances representing a broad spectrum of molecular mechanisms (Hu 1985).

The metabolic potential of the studied cell type can also influence the modifying effect of SCE-induction, as shown by experiments using indomethacin as a suppressor and benzo-<α>pyrene (BP) or dimethylbenz<α>anthracene (DMBA) as inducers of SCE (Abe 1986). Indomethacin moderately suppressed SCE induction by BP or DMBA in rat tumor cells (AH66B and R1), but it exerted no such effect in human C-HC-4 and Chinese hamster DON-6 cells. Similar findings on differing sensitivities to SCE induction of various cell types were also reported by the group of Speit. In contrast to earlier reports on the influence of SH compounds on the frequency of SCE in cell lines of the Chinese hamster, cysteine, cysteamine, and cystamine did not cause an increase in the SCE frequency in human lymphocyte cultures (Speit and Vogel 1982). Their comparative investigations showed that SCE induction by SH compounds is probably due to the inability of the Chinese hamster cells to degrade H_2O_2. They pointed out that the examination of a substance within one cell system does not necessarily permit a reliable statement on the DNA-damaging property of this substance.

A third group of substances with presumable SCE-reducing properties includes the vitamins A, C, and E (Table 15) which, as anticlastogens, were considered to act via their antioxidative properties.

A paradox situation was reported by Ockey (1981) who found that methylmethanesulfonate-induced SCEs are reduced by the BrdUrd used to visualize them if the DNA contained high levels of incorporated BrdUrd. This observation was substantiated by the finding of a dependence of the mitomycin-C-induced SCE rate on the amount of BrdUrd incorporated in DNA (Natarajan et al. 1986). As shown previously by Popescu et al (1980), the sequence in which MNNG and BrdUrd are added to cell cultures clearly affects the number of induced SCEs. When BrdUrd incorporation preceded MNNG treatment of V 79 cells, the average number of SCEs per metaphase was reduced to 70–80% of that found in cells which were treated with MNNG for 2 h prior to the addition of BrdUrd, and even to 60%, if three division cycles were considered. This fact should be kept in mind when seemingly contradictory data from experiments on the reduction of the SCE frequency by anticlastogenes are discussed.

The findings of a reduction due to cysteine (Abramovsky et al. 1978) and sodium selenite (Ray et al. 1978) in the SCE frequency induced by clastogens prompted a series of further studies on the anti-SCE action on human lymphocyte cultures (see Table 15).

A group of researchers (E.G.) carried out comparative studies on this test system using the sulfhydryl anticlastogens AET, HCT, and l-cysteine, but also vitamins C and E, and sodium fluoride as potential protectors; trenimon, the chelating agent 8-hydroxyquinoline sulfate, and S9-mix-activated cyclophosphamide were used as potent clastogens and SCE inducers. In the first test series, clastogen and protector were added to the cultures simultaneously 24 h prior to harvesting. None of the tested sulfhydryl compounds significantly decreased the SCE frequency induced by trenimon or 8-hydroxylquinoline sulfate, while the chromosome-breaking effect of both clastogens was found to be distinctly reduced by these anticlastogens in the same cells (Gebhart and Kappauf 1980).

In the second series, BrdUrd, at a final concentration of 10^{-5} M, was applied for the last 24 h of culture (i.e., together with the clastogen and anticlastogen) to allow an evaluation of SCE frequencies in those lymphocytes, which had performed two rounds of DNA replication in the presence of the test substances. The trenimon-induced SCE frequency, under these conditions, was only slightly higher than in the experiments reported before. Neither AET nor HCT could reduce this rate significantly.

The S9-mix experiments reported in Section 3.2.2.3 were also extended to the action of AET, VC, VE, and NaF on the SCE-producing effects of trenimon and cyclophosphamide. Here, the test substances were applied for only 2 h at a time when the majority of lymphocytes were presumed to be in the G1- or the early S-phase of their cell cycle. Similar to the data presented above, a more or less distinct decrease in the breakage rate induced by trenimon was observed with all anticlastogens without the addition of S9-mix. In the same experiments, however, only AET slightly reduced the SCE frequency caused by the clastogen, while the other anticlastogens apparently lacked any significant anti-SCE activity. In previous studies (Gebhart and Becher 1976) it could be shown that some anticlastogens act differently on diverse types of structural chromosome damage, particularly if open breaks vs interchanges were compared. Therefore, in the cited SCE study (Gebhart 1984), the action of the anticlastogens on the trenomin-induced interchange rate was compared with that on SCE. Striking differences between the reaction of both cytogenetic phenomena could be observed again in most cases, with the exception of AET, which decreased the frequency of both trenimon-induced interchange and SCE. The other protectors were only active in reducing the interchange rate in conventionally stained cells, but not the SCE rate in BrdUrd-labeled lymphocytes. If both cytogenetic phenomena were estimated from BrdUrd-labeled slides, the observed differences between both cytogenetic phenomena were less striking, but absolute numbers of interchanges were much lower in these metaphases than in BrdUrd-unlabeled ones.

Although under the influence of metabolizing S9-mix preparations the absolute numbers of SCEs induced by trenimon were distinctly increased, no clear-cut change in the action of the anticlastogens on the yield of SCE could be observed. The slight anti-SCE activity of AET reported for the experiments in which S9-mix was not added, was even decreased, and a very slight SCE-decreasing effect of vitamin E became apparent under these metabolizing conditions.

Experiments employing the mixed-function-oxidase activation-dependent cytostatic clastogen cyclophosphamide (CP) yielded less definite results, apparently due to a less favorable clastogen/anticlastogen relation in terms of molarity. Although a slight decrease in the yield of break events induced by 5×10^{-5} M CP could be observed when AET, VC, or NaF were applied as anticlastogens, SCE frequency caused by 10^{-5} M CP remained unchanged in these experiments. In addition, the respective studies comparing the effect of the anticlastogens on CP-induced chromatid interchanges with that on SCE also yielded no clear correlations between both phenomena.

The use of hepatoma cell lines for studying genotoxic effects induced by mutagens as well as their modification seemed to be an obvious technique to study the direct influence of metabolism without using the rather toxic S9 preparations (Grady et al. 1986).

Benzidine was found to induce SCE in these liver-derived cell lines in a dose-dependent fashion, but failed to induce SCE in two fibroblast lines with low metabolizing capability used for control. Indomethacin, a presumed inhibitor of this metabolic pathway, was highly effective in inhibiting induction of SCE by benzidine in the liver-derived cell lines.

It should be mentioned here that also a few reports on the SCE-reducing activity of anticlastogens from mammalian in vivo test systems have been published so far. For instance, Krishna et al. (1986), using ascorbic acid as an anticlastogen, observed a reduction in the number of SCEs induced by cyclophosphamide and mitomycin C in bone marrow and spleen cells of mice. At the highest concentration of the vitamin (6.68 g/kg) approximately 75 and 40% SCE inhibition in both cell types was noted for cyclophosphamide and mitomycin C, respectively. An influence of amino acid supplementation to unbalanced diets was found for the SCE frequency in the bone marrow of rats exposed to cyclophosphamide (Chen and Lin 1987). A low (23–27%) but significant reduction in the mitomycin-C-induced SCE rate was observed in bone marrow cells of mice treated with the immunopotentiators Kristin and Lentinan (Hasegawa et al. 1989). However, it should not be neglected that some of the anticlastogens themselves can also readily induce SCE, as shown for instance for vitamin C and sodium selenite in Chinese hamster cell cultures (Galloway and Painter 1979; Mac Rae and Stich 1979; Speit et al. 1980b), but not for vitamin C in Chinese hamster or mice in vivo (Speit et al. 1980b; Krishna et al. 1986). Moreover, Mac Rae and Stich (1979) also reported on an SCE-inducing potency of cysteine, cysteamine, and glutathione.

SCE studies, though not representing a direct approach to all forms of anticlastogenic action, as discussed above, may nevertheless contribute important information on the specific mechanisms of chromosome-damaging effects and their modification.

3.5 A Mainly Historical Side-Glance at Anticlastogens and Ionizing Radiation

Although it was the explicit purpose of this book to discuss the protective action of various agents on the clastogenic activity of chemical mutagens, their activity versus the effects of ionizing radiation should not be neglected. Investigations on anticlastogenic action versus ionizing radiation were not only the first studies in this field of experimental cytogenetics, but they also contributed the first valuable evidence necessary for the formulation of basic hypotheses on the mechanisms of anticlastogenic action. These studies date back to the 1950s when the anticlastogenic activity of radioprotectors, which were of high actuality at that time, was detected on plant chromosomes (Mikaelsen 1952, 1954; Riley 1952), and also in bone marrow cells of the mouse (Devik 1954).

Actually, the first decade of this research was governed by studies on submammalian test systems (see Rieger and Michaelis 1967 for detailed review). In most of these studies anaphase fragmentation or bridges, metaphase breaks, and translocations were the main parameters for estimating clastogenic, and subsequently, anticlastogenic activity in that period.

L-cysteine, cysteamine, cystine, glutathione, aminoethylisothiouronium (AET), thiourea, dimercaptopropanol, as well as several sulfites, sulfates, thiosulfates, and "natural" agents (sugars, amino acids, vitamins, some other naturally occurring acids, and indenes) were found to protect chromosomes from radiation-induced damage to various extents (see Gebhart 1974 for detailed information). "Dose reduction factors" (DRF) of about 2 were sometimes achieved applying these anticlastogens. While most of the experiments were performed on plant cell material (e.g., *Tradescantia, Allium, Vicia, Hordeum*), a few were also dedicated to insects (e.g., the grasshopper *Gesonula*, or the fruitfly *Drosophila*).

Although tests on mammalian cells were already performed from the first years of antimutagenesis research (Kaplan and Lyon 1953; Devik 1954; Devik and Lothe 1955), it took several years until a more systematic analysis could be extended to mammalian and human cell material. A series of tests was performed on human cell cultures (mainly peripheral lymphocytes). While cysteamine displayed a good protective effect against X-ray-induced chromosome damage in HeLa and human kidney cells (Firket and Mahieu 1966; Vos and Kalen 1968) as well as in all tests with mammalian cell cultures, no protective action of this substance could be found in human peripheral lymphocytes (Nasjleti and Spencer 1969). Without going into further detail, it should be mentioned that more recently the main concern has also focused on natural anticlastogens in the field of radiation anticlastogenesis. For example, chromosome aberrations induced by UV-light and X-rays in CHO cells were found to be suppressed by posttreatment with antimutagenic flavourings, such as anisaldehyde, cinnamaldehyde, coumarin, and vanillin (Sasaki et al. 1990a): "UV- or X-ray-irradiated surviving cells increased in the presence of each flavouring. X-ray-induced breakage-type and exchange-type chromosome aberrations were suppressed by the vanillin treatment in G1 phase of the cell cycle and a greater decrease in the number of X-ray-induced chromosome aberrations during G1 holding was observed in the presence of vanillin. Furthermore, a greater decrease in the number of X-ray-induced DNA single-strand breaks was observed in the presence of vanillin. Treatment with vanillin in the G_2 phase suppressed UV- and X-ray-induced breakage-type but not exchange-type chromosome aberrations. The suppression of breakage-type aberrations was assumed to be due to modification of the capability of the post-replicational repair of DNA double-strand breaks. These G_1- and G_2-dependent anticlastogenic effects were not observed in the presence of 2′,3′-dideoxythymidine, an inhibitor of DNA polymerase β. Based on these results, the anticlastogenic effect of vanillin was considered to be due to the promotion of the DNA rejoining process in which DNA polymerase β acts." This example was cited in more detail, not only because it is one of the most recent publications pertaining to the anticlastogenic effects of natural products on the action of ionizing radiation, but also because it demonstrates impressively that these studies, if performed sophisticatedly, also yield important evidence of the molecular mechanisms underlying the observed anticlastogenic effects.

It was quite remarkable that a distinct action of radioprotectors was not only observed in mammalian and human cells irradiated in vitro, but also after in vivo exposure to irradiation of test animals. As shown in Table 16, data obtained thus far exhibit great similarity to those obtained from studies on plant material.

Application of the protector before irradiation, in most cases, other than with exposure to chemical clastogens, was the most effective treatment mode. However, there were also several reports on the anticlastogenic efficacy of posttreatment (e.g., 1,4-dithiothreitol, spermidine, thyroxine, 2-deoxy-D-glucose; DNA, superoxide dismutase, catalase, erythromycin, streptomycin, dihydrostreptomycin, penicillin, interferon, cystaphos, gammaphos, linoleate) or simultaneous treatment (e.g., AET + ATP + serotonin, l-cysteine, α-tocopherol, dimethylsulfoxide, penicillin, several SH compounds).

Of particular interest were in vivo studies using germ cells as targets of protection against the clastogenic effect of ionizing radiation. These investigations were performed mostly in the "early" years of anticlastogen research in mammals. It, therefore, seems worthwhile to take a retrospective view of the respective findings. The complexity of the mammalian organism, due to the potential of various metabolic factors as well as to the different sensitivities of the various germ cell stages, is reflected by contradictory results. Lüning et al. (1961) reported that cysteamine protects mouse spermatozoa from the induction of dominant lethal mutations (which were later shown to reflect mainly chromosomal changes), whereas the studies of Kaplan and Lyon (1953) in mice yielded negative results. Gugushvili et al. (1972) were also able to show a decreasing effect of cysteamine on the induction of dominant lethal mutations in mice, but the protector did not affect the rate of induced reciprocal translocations in spermatogonia. AET also distinctly reduced the rate of the former type of mutations (Ehling 1964a,b), although Takeda and Sugahara (1960), after injecting AET into the scrotal sac of mice, Leonard and Maisin (1963), after treatment of spermatogonial stages, and Nuzhdin and Niznik (1963), after application of the same agent to spermatozoa, failed to observe any protection against induced lethal mutations. Ehling (1964b) was also able to demonstrate an antimutagenic activity of l-cysteine in the domi-

Table 16. Substances for which an anticlastogenic action was shown in experiments with ionizing radiation (X-rays, γ-rays)

Anticlastogenic effect detected in			
Plant cell systems	Mammalian cells systems (including human cells)		
	In vivo:		In vitro
	Germ cells	Somatic cells	
AET	AET	AET	AET
L-cysteine	L-cysteine	L-cysteine	L-cysteine
Cysteamine	Cysteamine	Cysteamine	Cysteamine
	Cystamine	Cystamine	Cystamine
Cystine			Cystine
Thiourea	Serotonin		Mercaptoethanol
Glutathione	Glutathione		Glutathione
	WR 2721	WR 2721	WR 2721
Dithiothreitol			Dithiothreitol
DMSO[a]		DMSO	DMSO
	Interferon[b]	Interferon	Interferon
	Penicillin[b]	Penicillin	
Sulfoguanidine		Linoleate	Linoleate
α-Tocopherol			α-Tocopherol
Spermin			Spermidine
Colchicin	Colchicin		
Dimercaptopropanol	Erythromycin	Thyroxine	Deox-D-glucose
	Streptomycin	DNA	DNA
Na-hydrosulfite	Dihydrostrep-	Serotonine	Antipain
	tomycin	Methoxytrypt-	Glycerol
Na-thioglycolate	Thiocarbamide	amine	Nicotinamide
Na-sulfhydrate	Cystaphos	Liv.52[a]	Unithiol
Na-thiosulfate	Gammaphos	Camphor	Superoxide
Na-hydrosulfate	WR 2721	Hydroxytrypt-	dismutase
Na-metabisulfite	Hydroxytrypt-	amine	Catalase
Na-pyrosulfate	amine		t-Butanol
Fructose			Methanol
Glutamic acid			Ethanol
Fumaric acid			Ethylene glycol
Maleic acid			Isopropanol
Sorbic acid			Indene F37
Gallic acid			Anisaldehyde
Cinnamic acid			Cinnamaldehyde
Vitamin A-palmitate			Coumarin
Vitamins B6, C			Vanillin
Asparagine			
Serine			
Cupferron			
N-etylmaleimide			
Indene F25	F38[b]		
Giberellic acid			
Hydroquinone			
Butylhydroxyanisol			
Butylhydroxytoluol			
Pyrogallol			
Kinetin			

[a]Abbreviations: DMSO, dimethylsulfoxide; Liv.52, mixture of herbal extracts.
[b]Studies performed on *Drosophila*.

nant lethal test in the mouse. Leonard and Deknudt (1972, 1973) studied chemical radioprotection against the induction of chromosome rearrangements in spermatogonia in mice. AET was ineffective in modifying the nature or rate of these rearrangements in diakinesis, i.e., the first metaphase stages of meiosis induced by 500 R if the irradiation was applied to spermatogonia. Pretreatment with a mixture of reduced glutathione (GSH), AET, mercaptoethylamine, cysteine, and serotonin creatinine sulfate (5-hydroxytryptamine), however, decreased the percentage of abnormal metaphases significantly. After irradiation of mouse spermatogonia a reduction in the frequency of chromosomal rearrangements was found in dividing spermatocytes if cysteamine as well as AET were applied before irradiation (Savkovic and Pecevski 1973; Bajrakova et al. 1974). The same protector was later found to also decrease the frequency of micronuclei in bone marrow cells of mice exposed to gamma radiation (Garriott and Crowe 1983) and displayed antimutagenic properties in mixtures with serotonin and ATP with regard to X-ray-induced reciprocal translocations in mouse spermatogonia (Benova 1986).

Without any claim to completeness these selected examples may suffice to demonstrate the close relationship of radio- and chemoprotection in in vivo mammalian systems.

One of the most important sources of our knowledge on the molecular mechanisms of anticlastogenic activity, once again, is radiobiology. Since many recent investigators in this field may no longer be aware of the historical roots of this evidence, the opportunity should be taken here to briefly review some of the historical concepts originating from this field of research.

Again, many of the pertinent studies date back to the early years of anticlastogen research and, being of some relevance also to our central considerations on anticlastogenic action versus chemical mutagens, they will be regarded in this context. Some of the well-known radioprotectors act via random scavenging of radiation-induced radicals in the solution or the suspending medium. Thus, the protection of DNA by the aminoalkylthiosulfates and thiophosphates may be due to radical scavenging (Shapiro and Kollmann 1969). This mechanism of action may also be part of the protective activity of thiols and disulfides. In present terms, this form of extracellular radical scavenging would be characterized as a "desclastogenic" mechanism.

Data of Roots and Okada (1972) on the efficacy of scavengers suggested: (1) that the protectable fraction is mostly of indirect action; (2) that the OH radical plays a major role in radiation-induced, single-strand DNA breaks in cultured mammalian cells; (3) that a major protection mechanism of indirect action is radical scavenging or a competitive reaction; and (4) that the protection mechanism by SH compounds may consist of radical scavenging as well as some other type of reaction resulting in radiation protection. Modig et al. (1971) reported on the release of thiols from mixed disulfides induced by cysteamine which enhanced the radioresistance of cells. The cells apparently contain a "defense mechanism" consisting of GSH and glutathione peroxidase, which can rapidly destroy deleterious peroxides. The efficiency of this mechanism can be enhanced by increasing the intracellular concentration of the GSH (Revesz 1969).

Localized radical scavenging at the surface of the target molecule requires a reaction of the protectors with DNA. This was first demonstrated by Jellum (1965, 1966) who showed that diaminodisulfides protect DNA because they contain a disulfide and are localized on the DNA. They are bound through their amino groups to phosphate groups of the nucleic acids. It was assumed that this reversible binding protects the DNA molecule when irradiated in vitro (Milvy and Pullman 1968). A very similar binding to DNA of polyamines was reported by Johnson and Bach 1967.

Nucleohistones contain reactive thiol and disulfide groups which readily form mixed disulfides with cystamine as well as cysteamine (Jellum and Eldjarn 1965; Eldjarn and Jellum 1969). Kollmann et al. (1967) examined the effect of the presence of disulfides on the viscosity of salmon sperm DNA and found that the efficacy of guanidoethyldisulfide presumably was produced by radical scavenging at the DNA surface.

These mechanisms of reactions of protectors with DNA were mainly found in the study of the anticlastogenic action of protectors on genetic damage induced by ionizing radiation. In the meantime they have been basically completed or in some cases even replaced by modern evidence provided by biochemical and molecular genetic investigations. Nevertheless, it was tempting to present a historical view for comparisons with the present state of knowledge, which has recently been reviewed by de Flora and Ramel (1988), and, to point out specific mechanisms, e.g., the protective role of glutathione and glutathione transferase by Ketterer (1988), the nature of oxidants and antioxidants by Hochstein and Atallah (1988), and the mechanisms of inhibition of free-radical processes by Simic (1988).

4 Practical Consequences

The application of the findings of anticlastogen research discussed above, in principle, could be of practical importance. First, their consequences for anticlastogen research itself have to be considered, including the possibilities and limitations of future test strategies; and, second, the consequences of the findings of anticlastogenesis research for human risk populations must be discussed. This must also include considerations on the compatibility of anticlastogens and the knowledge gained from their practical application in other (e.g., therapy) fields.

4.1 Practical Consequences in the Field of Anticlastogen Research

Several of the substances found to be potential anticlastogens are also components of the cell culture media used for growing the target organisms of cytogenetic, in vitro tests. This, however, means that every change in the composition of these media may fundamentally influence the outcome of the subsequent experiments. A comparison of the content of SH compounds of several standard media clearly documents the dimensions of this influence (Fig. 12).

In addition, although even less defined, the content of natural anticlastogens of the usually applied sera is of at least equal importance for the outcome of in vitro experiments, as clearly shown by Baret and Emerit (1983). However, it would not be incorrect to also assume variations in the contents of antimutagens in the daily food of experimental animals (as shown, e.g., for butylhydroxyanisole and ethoxyquine by Bueding et al. 1986). Although through steady standardization these contents can be adjusted, it cannot be absolutely excluded that many of the seemingly contradictory results from experimental clastogenesis and anticlastogenesis studies might be due to differing amounts of antimutagenic factors of the used substrates. As documented by data from anticlastogen research, the endpoints of mutagenic action affected by the respective protectors are of crucial importance for the conclusions drawn from the experiments. The classical cytogenetic approach is based on the evaluation of structural chromosome damage of various types.

Depending on the aim of each respective study, these types of aberrations may be of different importance. As, however, some of the anticlastogens display different levels of protection on these various aberration types (cf. Fig. 2), this fact must be taken into consideration when planning studies on anticlastogenic action.

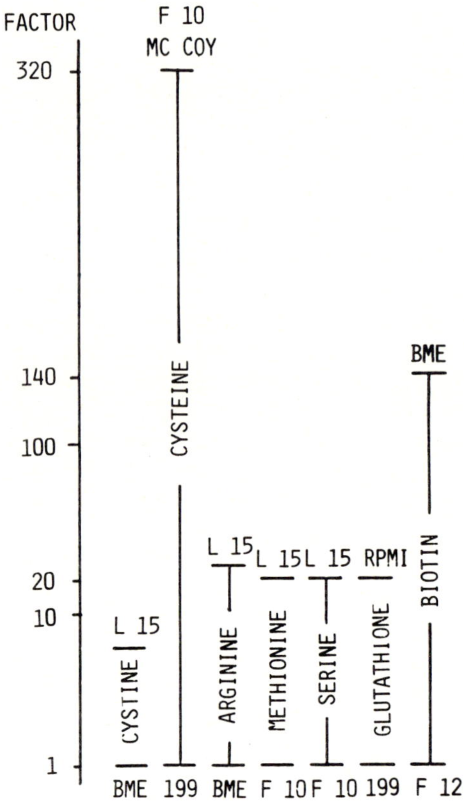

Fig. 12. Comparison of the contents of potential anticlastogens in various cell culture media in common use (according to catalogues of the producers)

With regards to the specific estimation of anticlastogenic effects, new possibilities were discussed on the basis of micronuclei analysis in human cells, especially oral mucosa cells and lymphocytes. The micronuclei method is easier to apply than the analysis of aberrations and SCE and very often is just as precise. It is quite reasonable to apply this method on a large scale in risk groups when studying the effects of anticlastogens. Krepinsky and Heddle (1983) emphasized that "although the micronucleus test can be regarded as a substitute for direct chromosomal aberration analysis, it is only complimentary to the other mutagen sensitivity tests such as cell survival or sister chromatid exchange." Of course, it cannot be recommended to eliminate the analysis of chromosomal aberrations or SCE from the complex of genetic monitoring of antimutagenic effects. However, only the analysis of all mentioned parameters in the different risk groups can provide a real basis for future investigations.

Another essential aspect is that the practical interpretation of the data from anticlastogenesis experiments should include the correct test design in the in vitro and/or in vivo systems, and the correct estimation of the obtained results. The effects of testing the modifying potential of chemicals are equally important as those revealing mutagenic activity. An excellent example is caffeine, which, depending on the used test system and its application, displayed comutagenic or even antimutagenic activities. Very often, at the beginning of the respective research, the investigator does not know whether a substance is a sensitizer or a protector. Thus, the experimental design should be able to consider two-sided results.

The methodology of investigation and use of anticlastogens are directly connected with methods to extrapolate data regarding chemical mutagenic action in different test systems. In the early stages of anticlastogen investigations, excessive and overenthusiastic extrapolations of obtained data were made, e.g., from plant cells to groups of exposed workers. Now the approach is more sophisticated and quantitative, but the main idea remains the same, i.e., to use the data obtained from various model systems to estimate the protective effects on the genetic risks of certain groups of exposed individuals. These conclusions, however, have now been based on substantial experience from genetic toxicology, as shown above.

One of the most universal questions of anticlastogenesis research is, what substances should be investigated for their anticlastogenic properties? The answer depends on the goal of the research. Either it is necessary to estimate the modifying properties of a new, unknown substance, and then research can begin from the analysis of literature or data banks on the properties of related compounds; or the substance is well known and is used as a tool to understand the mechanisms of modification, individual variability in response to its administration, or for extrapolation reasons, etc. Therefore, it is recommended to use anticlastogens not only from one particular class, but, for reasons of comparison, from several classes of chemicals with different properties or mechanisms of inhibition of mutagenesis, if possible. Very often, the detected action represents just the local interaction of the substance with the mutagen, which is very difficult to interpret, especially if quantitative estimates of the effect are not considered. For the possible application in groups with genetic risk, it is of course, more reasonable, under otherwise equal conditions, to prefer the protectors of natural origin, which are physiologically better tolerated than synthetic ones might be. Which mutagens can be recommended as standard test substances to study the modification of their effects? Only those mutagens whose action has been qualitatively and quantitatively well characterized seem to be sufficiently suited. It is certainly reasonable to use the same principles noted above. If possible, it would be very productive to compare the anticlastogenic properties and the action of different mutagens from the different classes. Here, two main principles of decision-making can be used. (1) The decision can be based on the type of molecular reaction, e.g., alkylating agents and antimetabolites or alkaloids, if that choice or grouping proceeds from the logics of research. (2) It can be based on the differences in chemical structure in one class of mutagens, e.g., "one-centered" and "multicentered" alkylating agents. As mentioned earlier for the protectors, if the choice does not contradict the logic of the respective study, it seems to be more reasonable to use those mutagens that are also used in medicine. What concentrations of the anticlastogens should be tested? For all investigations it is necessary to delineate the tolerated anticlastogen concentrations by preliminary experiments without mutagen. Then having defined the subtoxic concentration range, when there is no reliable data available from the literature, the applied dose can be decreased in small, equal intervals in in vivo tests and in larger intervals in in vitro systems, where the concentration of anticlastogen does not yield a linear dose dependence of the protective effect. In the experiments discussed above, usually one or two concentrations of anticlastogen were applied that gave a stable protective effect with the used mutagens.

Regarding the optimal experimental conditions for testing the protective effect in human cell cultures, e.g., lymphocyte cultures, it is important to use the same culture media in all experiments of any series. For example, it was shown that the use of RPMI 1640 or Eagles culture media resulted in a few hours differences in the onset of the total first and second divisions, the former (RPMI) causing a faster proliferation. The same was also found for other components of the culture, e.g., serum, PHA, etc. The use of blood obtained from the same donors appears to be less complicated. However, if no significant difference in the response to the mutagen dosis exists among the performed experiments, then it is also possible to use different donors to study chromosomal aberrations. For the analysis of induction and modification of SCE and cell proliferation, special statistical approaches can be used to overcome the differences caused by the interindividual variability in the speed of proliferation or the proportions of lymphocytes in different divisions. For such studies it is also preferable to use blood obtained from the same donor, which has no peculiarities regarding lymphocyte proliferation, or which does not differ significantly from the results of parallel studies carried out on the blood probes from only a few donors.

A problem which needs to be urgently solved is the question if, and to what extent, the enzymatic makeup of erythrocytes can modify the outcome of anticlastogen studies, if -- as usual -- whole blood cultures are used for the tests. Of course, these erythrocytes are present in all experimental setups (i.e., positive as well as negative controls, and all cultures and exposed to clastogen and anticlastogen). Nevertheless, they could reveal varying influences on the various experimental variants.

The visible effect of anticlastogens depends on the level of aberrations, seen in Table 17, which is based on the Poisson distribution. Although the data are analogous to those of Šrám and Kuleshov (1980), which were developed for sample sizes in order to measure the increase in aberration frequency, the present table is reciprocal to their figures. In Table 17 the level of aberrant metaphases (2.34) is twice the level of the spontaneous frequency of aberrant metaphases obtained in the Laboratory of Mutagenesis of the Institute of Medical Genetics (Academy of Medical Sciences of the USSR). The sample size used in Table 17 is based on the analysis of 100 metaphases from each individual and each variant of culture, which was treated with the protector. The protective effect in terms of percentage of decrease in the levels of aberrant metaphases must be equal to 25–50% in order to investigate realistic sample sizes. Using sample sizes that are too large can cause a great bias due to intrasample variation, including contact with the mutagen, individual sensitivities, etc.

In order to elucidate the nature of the processes that favor the decrease in intensity of the mutation process, it is necessary to investigate models which describe the protective effect. Mechanisms of action of chemical mutagens in

Table 17. Sample sizes necessary to detect a decrease in the level of aberrant metaphases (AM) for different effectiveness of protective action (type I error, $\alpha = 0.05$)

Effect induced by mutagen in % of AM	Decrease in AM induced by protective action (%)	Sample size (units)[a] $\beta = 0.05$	$\beta = 0.10$
20	50	2	1
20	25	8	6
20	10	51	40
15	50	2	2
15	25	10	8
15	10	68	54
10	50	3	2
10	25	16	12
10	10	103	81
5	50	6	5
5	25	30	23
5	10	205	163
2.34	50	13	10
2.34	25	64	50
2.34	10	438	345

[a]Unit = 100 metaphases; β = type II error.

human cells are being investigated only now, and because of this lack of complete understanding, the elucidation of the mechanisms of their modification is further complicated. However, sometimes even a description of alternative quantitative processes connected with cytogenetic parameters is sufficient to understand some of the possible pathways of protective action. Our approach is based on the use of diagrams (blockschemes), which detect different ways of protective action. The method of their construction was presented by Shannon (1975). Using such schemes may be a productive way to understand the comparative role of the anticlastogenic mechanisms. It seems reasonable to apply at every stage of research a multimethodological approach, using contrasting test systems or alternative models in each case, for example, cell lines with high and low repair activity, different periods of anticlastogen or mutagen treatment, or different times of culture fixation. The obtained results show which alternative is less suited to explain the mechanism providing the protective effect. Thus, by rejecting the least possible explanation for the effect, we can obtain stepwise valuable information. To understand the mechanisms of anticlastogenic effects, it may be simpler to choose between the contradictory theoretical models representing the nature of cytogenetic changes and the role of metabolism or repair mechanisms in their modification.

For example, several diagrams are presented (Figs. 13–16), from dozens of possible decision-making processes, that can be used (and were used by us) to analyze protective action. From these, investigators can construct their own variants.

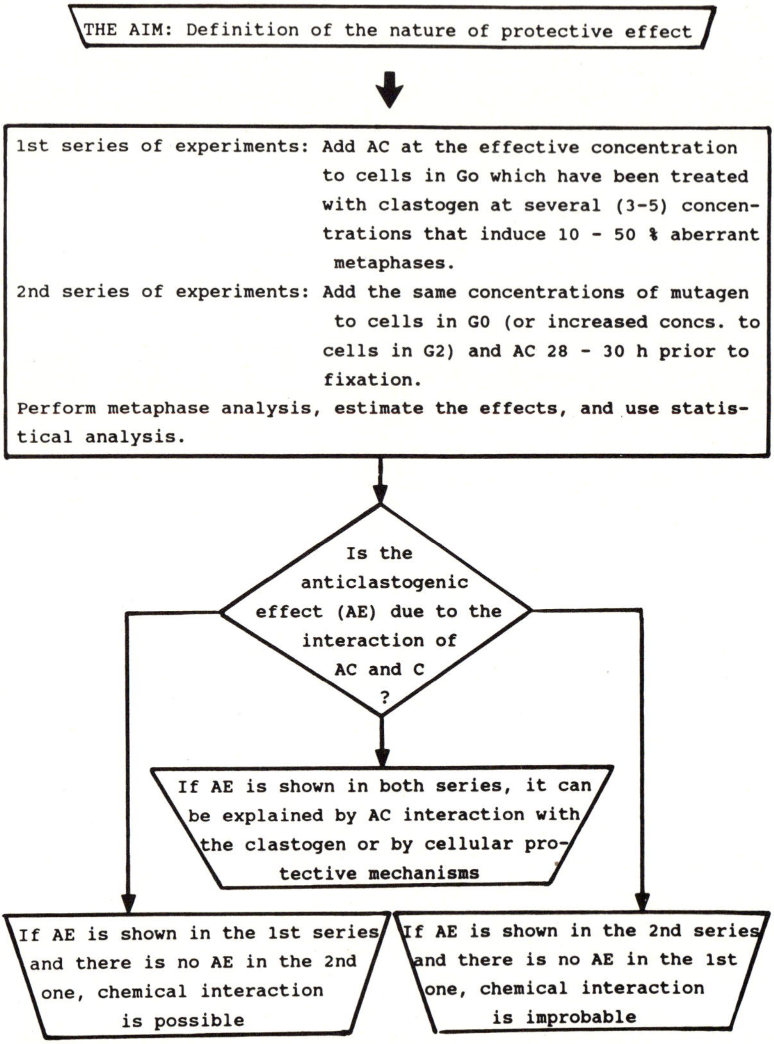

Fig. 13. Diagram presenting different ways of analyzing anticlastogen action

The decision-making process for a possible explanation to the protective effect is presented in Fig. 13. Considering the possible chemical interaction, a choice can be made between: (1) the possible chemical interaction between mutagen and anticlastogen and (2) the low possibility of that interaction and probable real antimutagenic effect. According to the classification of de Flora and Ramel (1988) these variants can belong to: (1) inhibitors of mutagenesis acting extracellularly and deactivating mutagens by chemical reaction, and (2) to the inhibitors of mutagenesis acting intracellularly, and among them, to the class of modulators of DNA replication or repair, or as the authors noted: "bioantimutagens or anti-

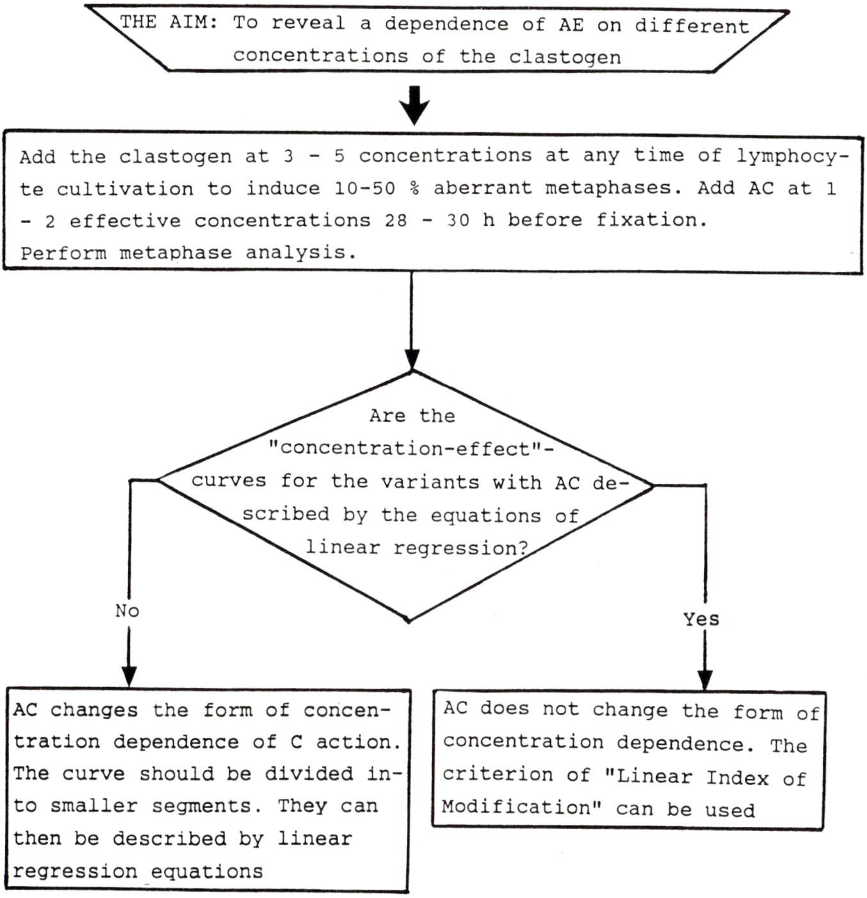

Fig. 14. Diagram of the analysis of the dependence of the anticlastogen effect on the mutagen concentration

mutagens in strict sense according to Kada et al. (1982)." In Sections 3.2.2.1 and 3.2.2.2, every experimental table presents the choice between the presented higher variants, and each time, the real antimutagenic effect was noted.

The artifact created by the selective displacement of cells can play an important role in the reduced cytogenetic effect, as shown for thio-TEPA and WR 2721 by Wolff and Arutyunyan (1979). The artifact of cell selection caused part of the protective effect. This was revealed only by quantitative comparison of M1, M2, and M3 metaphases visualized by BrdUrd labeling (Fig. 15), which showed the nature of the changes. Details on the proportions of cells from different divisions in the experiments with anticlastogens can be found in Section 3.2.2.1.

Figure 14 shows a practical approach to the analysis of the anticlastogen effect on the "concentration-effect" curve of the mutagen. It was constructed according to the mathematical model of Chebotarev and Yakovenko (1974) for dose-effect

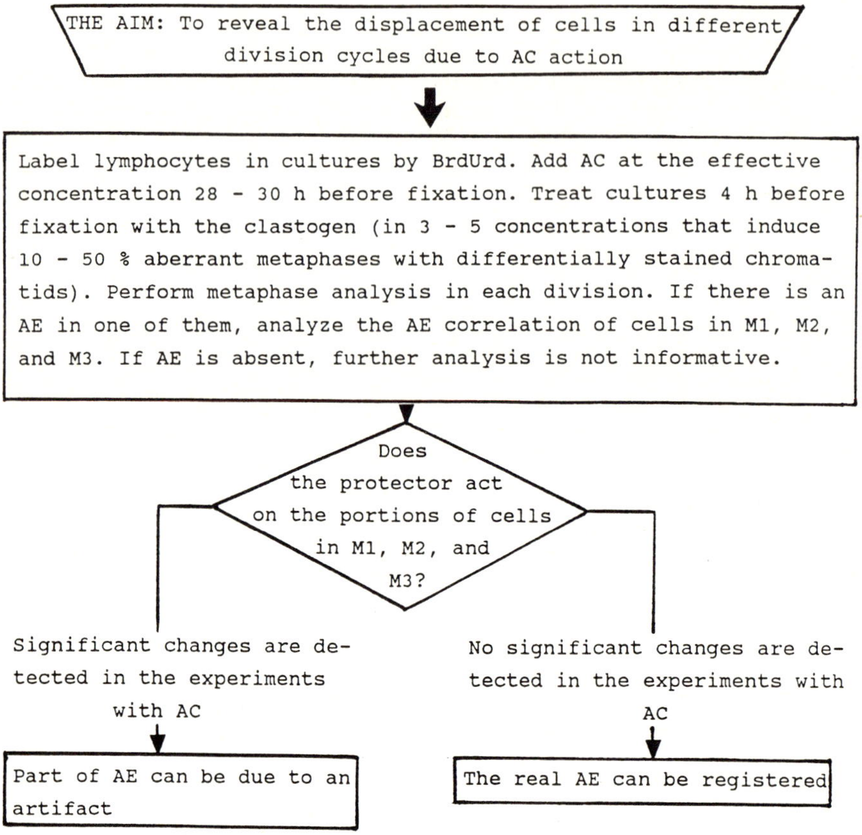

Fig. 15. Diagram presenting the analysis of possible cell displacement by the anticlastogen in human lymphocyte cultures using BrdUrd labeling

dependences (see Chap. 2) and on the basis of the universal estimate of the modifier's effect, i.e., the linear index of modification (LIM). The latter was based on the important condition of the approximation of the dose-effect curves for mutagens (with and without modifier) by the linear regression equations.

The same approach can also be used for sensitizers. In most of our experiments the variants with mutagens and anticlastogen do not significantly deviate from the linear regression which made the application of the LIM possible.

The approach with anticlastogens, in addition, necessitates the total screening of all prospective substances on one level of the experimental test systems. The application of different tests is inevitable. The first possible and very necessary test, i.e., the microbial test, is not considered here, even when such studies deal with antimutagens. However, the second level, i.e., the application of animal cells, and the third level, risk assessment (Fig. 16), are very important because they deal with the systems under discussion. If some modifications are introduced in the

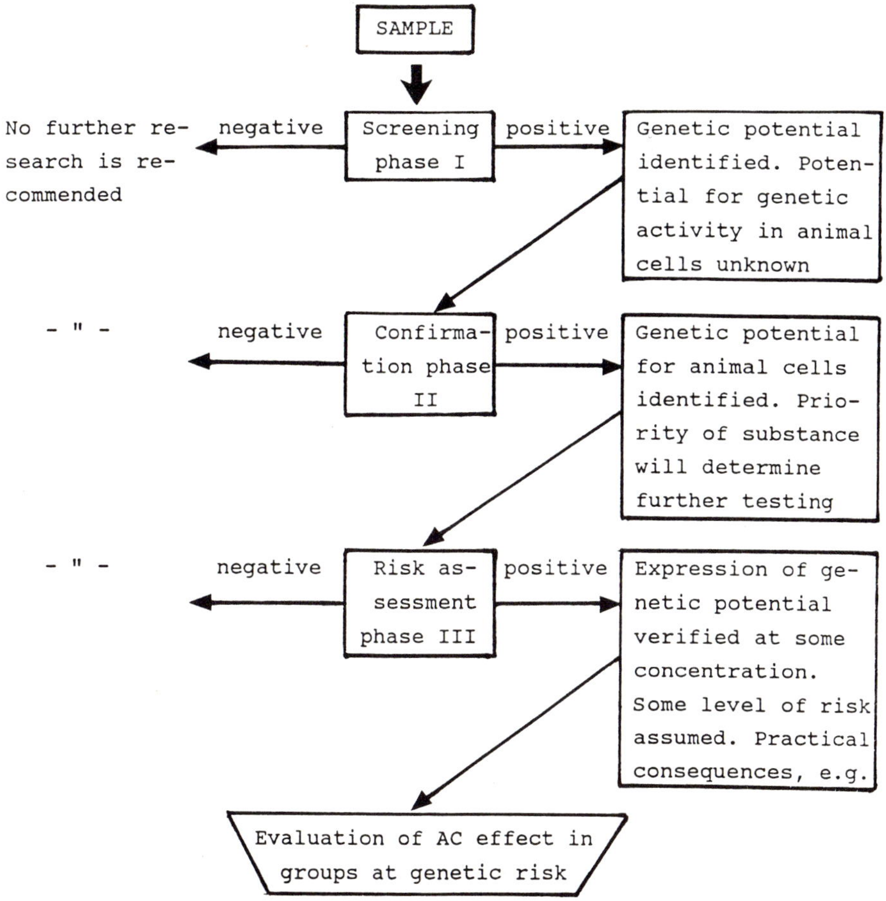

Fig. 16. Tier approach to anticlastogen testing (After Brusick 1980)

schematic representation of the tier approach (Brusick 1980), we can use it not only to estimate mutagenic, but also antimutagenic action.

4.2 Practical Consequences for Humans

Not only the consequences for experimental anticlastogen research are important, but also the practical meaning of anticlastogenic action for individuals exposed to mutagens in the framework of a modern industrial environment.

With this in mind, one must distinguish between the views of general genetics regarding the protection of human germ cells from the induction of heritable changes, on the one hand, and the aspect of protecting somatic cells against mutagenic/carcinogenic effects, on the other. Obviously, it is not the purpose of practical antimutagen research to decrease the spontaneous mutation rate of whole mankind or whole populations. This is provided by natural systems of protection, e.g., bioantimutagens, repair systems, and mechanisms of selection. The validity of this latter statement is supported by the result of extensive human genetic studies on the offspring of the atomic bomb victims of Hiroshima and Nagasaki (Schull et al. 1981; Hamilton 1982) which have not thus far been able to detect a decisive increase in genetic diseases, and also by the clinical studies on the offspring of patients, who underwent successful cytostatic therapy (Holmes and Holmes 1978; Li et al. 1979).

The development of antimutagens for practical use in limited groups of individuals with a high risk of exposure to mutagens remains the main and explicit aim of antimutagen/anticlastogen research. As shown above, the basis for progress in this field is limited by the lack of experience from in vivo test systems or observations on human individuals. Although the data from in vitro studies have clearly proven the existence of antimutagenic/anticlastogenic action, the way to practical application in man is still stony. As shown in Section 3.3, there are very promising approaches which will yield data of high practical importance in the future.

The production of exhaustive lists of desmutagens and/or bioantimutagens active in bacteria, which has come into fashion over the last years (see, e.g., International Conferences on Antimutagenesis and Anticarcinogenesis), does not answer the question of the practical feasibility and efficiency of antimutagens in humans. Therefore, definite intensification of in vivo research on antimutagens and -- depending on the suitable systems -- particularly of anticlastogen research is crucially needed at the present. Nonetheless, all the data collected from model investigations are of inestimable value for providing guidelines for further research. Possible candidate antimutagens are presented in Table 18. Until respective data have been obtained, it must be admitted that at the present applicable antimutagens for protecting human germ cells against mutagenic damage are not available.

With respect to the prevention of the formation of mutations in somatic cells, the situation seems to be more favorable, as this problem has been studied for

Table 18. Selection of "candidate" anticlastogens which seem to be suited for further intense (e.g. in vivo) testing

N-acetylcysteine	Vitamins C and E
N-acetylhomocysteinethiolactone	β-Carotene
WR 2721	Selenite
Polyamines	Difluoromethylornithine
Selenium	Propylgallate
Butylhydroxyanisole	Riboflavin
	Antimutagenic plant extracts
	Interferon ?

several years in a two-fold fashion. Parallel to the search for antimutagens/anticlastogens in somatic cells of mammals and man, intense research was also dedicated to anticarcinogenic action on suitable in vivo test systems of several of these substances. Our knowledge of the molecular mechanisms of malignant transformation is much greater today compared to 10–15 years ago. Therefore, it is certain that the diminishing number of mutation events in somatic cells will also reduce the number of events leading to malignant transformation.

Without going into details of the modern molecular theories of the processes leading to malignant transformation (see Sect. 1 for references), it can now be stated that the activation of protooncogenes by mutational mechanisms (including chromosomal rearrangements), the inactivation of tumor suppressor genes ("antioncogenes") by point mutations, chromosome deletions and rearrangements, or chromosome loss are the main causative factors which transform a normal cell into a malignant one.

The wide spectrum of mutations causing malignant transformation, on the one hand, and our knowledge presented above on the various effects potential antimutagens display on the various types of mutations, on the other, allow us to assume that no unique mechanism of anticarcinogenesis can be expected. It is insignificant for our considerations whether the respective mutational step is to be regarded as a step of initiation or whether a series of initial steps combined with promotional events are necessary for the malignant transformation. As mutational events are involved in all these steps, the prevention of mutations, in a general sense, should also be connected with the prevention of transforming mutations. These considerations are additionally supported by a series of parallel findings from antimutagenesis and anticarcinogenesis research. For example, some substances combining both features are presented in Table 19. Nevertheless, the

Table 19. Substances which were experimentally shown to act antimutagenically and anticarcinogenically

Abundant data available for	Data from single experiments available for
L-cysteine	Benzylisothiocyanate
Cysteamine	Phenetylisothiocyanates
N-acetylcysteine	Apigenine
Glutathione	Arginine
Vitamins C and E	Disulfiram
β-Carotene	Linolenic acid
Butylhydroxyanisole	Propylgallate
Butylhydroxytoluene	Bioflavonoids
Ethoxyquine	Thiosulfate
Retinoids	WR 2721
Selenium	
Polyamines	
Superoxide dismutase	
Catalase	

mechanisms of protooncogene activation, at present, urge us to concentrate on a more differentiated analysis of those activities. It appears rather simple to define an anticarcinogenic action as the reduction of the number of tumors induced by a certain carcinogen in a well-defined model system by an anticarcinogen. In contrast, due to the broad spectrum of mutational events caused by carcinogens, the definition of antimutagen seems to be more complicated, particularly, if the various endpoints of mutagenesis are influenced by it to a different degree.

As shown above, some of the findings on anticlastogenic action obtained in the 1970s have shown that several substances, which are able to reduce the number of open chromosomes and chromatid breaks, can increase the number of induced rearrangements. The latter, however, are undoubtedly anomalies of greater importance, for instance, with respect to the activation of cellular oncogenes, than the breaks, which are in most cases even lethal to the damaged cell. Therefore, antirecombinogenic substances should be differentiated from pure "antibreak" substances by using suitable experimental systems. Recalling these considerations, it is clear how difficult it is to judge the practical applicability of antimutagens/anticarcinogens without detailed knowledge of their mechanisms and spectra of action. The situation is additionally complicated by the fact that several agents, which have been shown to act by reducing the clastogenic action of mutagens, can also act clastogenically to a certain degree. Thus, the number of anticlastogens suitable for practical application is even more limited than believed earlier.

Not only the necessity of being metabolized, but also the great differences in the formed metabolites are further factors limiting the practical use of anticlastogens in man. This has been demonstrated by findings from anticarcinogen research which, for natural reasons, have primarily been based on in vivo experiments, and, which now include clinical investigations (e.g., Hanck 1985; Bertram et al. 1987; Boone et al. 1990).

These substances, which have been successfully used as therapeutics for some time, can be regarded as sufficiently compatible and well-analyzed with respect to their biological efficiency in humans (e.g., N-acetylcysteine: Gurtoo et al. 1983; Miller and Rumack 1983; N-acetyl-homocysteine-thiolactone: Stosiek and Gebhart 1974).

Others have been at least clinically tested regarding their compatibility (e.g., vitamins C and E, β-carotene: Bertram et al. 1987; Chen et al. 1988; McKeown-Eyssen et al. 1988; Boone et al. 1990; WR 2721: Phillips and Wasserman 1984; Turrisi et al. 1986) and others again are apparently harmless since they are components of natural food (vitamins, amino acids, fatty acids, polyamines, bioflavonoids).

If a substance, which has been defined as an antimutagen by an in vitro test on mammalian cells or microorganisms, is metabolically altered in higher organisms in such a way that it probably deletes its active site before it can reach the cell structures to be protected, then practical evidence will counter-indicate its use as an antimutagenic/anticarcinogenic therapeutic. For example, this is valid for substances carrying free SH-groups which can easily be dehydrated to a disulfide group, as shown, for instance, for cysteine. Although, in the latter case, the cystine molecule formed by this reaction, has been reported in a few cases to also act antimutagenically, most observations indicate a particular antimutagenic action of the intact SH-groups in the cellular milieu. Therefore, attempts have been made to "mask" these groups in such way that they are immediately reactivated within the respective target cells. An example for the therapeutic application of substances constructed in this way is the active component of the liver therapeutic and radioprotector, Reducdyn, N-acetylhomocysteine-thiolactone. The SH-group responsible for its observed protective action is released from the thiolactone ring only within the cellular milieu. Also, in the case of N-acetylcysteine, the acetyl group seems to play the role of a stabilizer of the SH-group, thus leading to an evident therapeutic effect in clinical tests (Braun et al. 1957; Tosch 1960). In this context it may be of interest that diallylsulfide, a flavor component of garlic, although not a sulfhydryl compound itself, can be converted to such in animals (Hayatsu et al. 1988). Although as yet no data on its anticlastogenic effect are available, its administration to mice effectively decreases carcinogen-induced colorectal adenocarcinoma (Wargovich 1987). As reported by Hartman and Shankel (1990), its anticarcinogenic activity was also demonstrated in several other test systems.

Organ-specific anticarcinogenic/antimutagenic efficiency, however, must not be neglected. Such an effect, for instance, has been found by Ito et al. (1985). While butylhydroxyanisole apparently protected the liver of the test animals from the induction of foci by nitrosamines, the tumorigenic action of nitroso compounds in the urinary bladder or the prestomach of the mouse was increased by the same compound. Also, ethoxyquine prevented the formation of liver nodes in these experiments, but increased the tumor formation in kidneys (Manson et al. 1987). As pointed out by de Flora and Ramel (1988), similar findings were also reported for other modifiers, for instance, ethanol and disulfiram, which change the organotropism of the carcinogenic effects produced by nitrosamines (Schmähl 1976; Bartsch et al. 1986). In addition, some of the antioxidants previously classified as anticarcinogens can act as tumor promoters if applied after a carcinogen (see Kahl 1986). These data impressively demonstrate that it is not sufficient to analyze anticlastogenic actions by using only a single test system or only one type of application to be able to reasonably judge the antimutagenic/anticlastogenic action of a given substance. This must be kept in mind particularly if reports of antimutagenic efficiency of fruit juices, vegetable cocktails, or other complex mixtures, which have been studied by the Ames assay only, are made the basis of fundamental recommendations for certain life-styles. Epidemiological studies provide convincing evidence of their protecting activity, but this must be proven further by various test systems and regarding various genetic endpoints.

Although no sufficient uniformity of results has been reached thus far to justify a general therapeutic application of anticlastogens, it should not be overlooked that a series of clinical and preclinical studies have yielded important evidence indicating the protective activity of some of these agents on the toxic side effects of cytostatic therapies without affecting the desired effect of the latter (Kline et al. 1973). This is of practical importance as this evidence could facilitate the practical application of antimutagens in such risk populations.

5 Concluding Remarks

Anticlastogenesis is an essential part of antimutagenesis. Basic and applied research on anticlastogens has not only revealed very valuable evidence of the mechanisms active in the induction of chromosomal aberrations by environmental mutagens and the principles of its modification, but it has also contributed effective concepts on the practical employment of this knowledge to the protection of individuals at risk. Particularly the findings on the anticlastogenic potential of "natural" substances (e.g., vitamins, amino acids, fatty acids, various plant antimutagens, and others) have essentially expedited these ideas. The problems of dietary inhibition of mutagenesis and carcinogenesis became one of the most important directions of research in the 1980s. The public interest has increasingly focused on these problems, especially since the works of Sugimura (1982), Ames (1983, 1984, 1986), and Wattenberg (1983). Not only large international conferences on the mechanisms of antimutagenesis and anticarcinogenesis in 1985 and 1988 (and most recently 1991) with the subsequent publication of the proceedings by Shankel et al. (1986) and Kuroda et al. (1990), a special issue of Mutation Research edited by de Flora (1988), but also many recent review publications reflect the intense research and interest in this area. Apparently, some of the dietary antimutagens (anticlastogens) show clear associations to epidemiological factors, not only on the basis of their general nature but also due to the very subtle peculiarities of their chemical structure, as pointed out recently by Renner (1990a,b)

Important stimulations came from the exploration of the process of malignant transformation which had shown that cancer is ultimately generated by mutational processes in somatic cells. Thus, beyond genetic protection of future generations by prohibiting germ cells from being damaged by mutagens, evidence increased that also somatic cells of exposed individuals must be protected from the detrimental effects of mutagens on their genetic material (Gebhart 1989).

Encouraging findings of a good tolerance of "therapeutically" applied antimutagens (see Chap.4) have additionally advanced practical anticlastogen research. In this context, the possibility should be mentioned that particularly natural anticlastogens could act as anticarcinogens not only by directly suppressing the mutational processes responsible for malignant transformation, but also by stimulating favorable attitudes in tumor patients by "strengthening" the natural forces of resistance (as apparently excess vitamins do; Bertram et al. 1987; Boone et al. 1990). In addition, one can speculate that particularly those antimutagens, for which a reducing effect on spontaneous mutation rates was shown, might suppress or delete mutationally damaged and, therefore, particularly aggressive tumor subpopulations, which could ultimately threaten the lives of patients.

The application of anticlastogens is not the only method of protection from environmental mutagens. There are many theoretical approaches concerning protection from increased levels of mutagenic effects. Alekperov (1984) discussed three different ways of prophylaxis versus genetic effects induced by environmental pollution.

The first way is at the technological level, i.e., the isolation of mutagenic products on the basis of technological changes in their production. The second way, at the component level, is based on the genetic-toxicological testing of many thousands of factors and the following replacement — where possible — of mutagenic factors by nonmutagenic ones. However, both approaches are difficult to realize. The third approach is the compensatory approach, which is based on the increase in the stability of the genetic apparatus of organisms and the activity of their repair systems, which depends on application of antimutagens.

Moreover, the usefulness of anticlastogens must be evaluated. The following questions should be considered: How reasonable is the use of protectors? Which anticlastogens are suitable, and when and where are they to be used? It was shown by Šrám and Kuleshov (1980) that only a very small part of all investigated risk groups has an average level of aberrant metaphases higher than 3%. Therefore, it is debatable whether it is reasonable to decrease such a level with chemicals that must be given permanently. It seems, however, to be very realistic and rational to use anticlastogens as an additive in oncologic therapies in order to decrease the levels of aberrations induced by chemotherapy, because the chronic use of protectors is unnecessary in these cases and the levels of aberrations to be diminished are higher.

It should also be taken into account that there are not only economical or statistical reasons for not treating exceedingly large groups of the population with the most effective antimutagens, even when the protective effects can be estimated. Due to such practical reasons, specialists in occupational medicine distinguish between risk groups, subgroups of "higher risk", and finally to individual risk estimates (Schulte et al. 1986, cited from Forni and Bertazzi 1987).

Of course, this does not mean that population prognosis, the epidemiological approach, or even environmental policy are less important in judging the problems of antimutagen application. There are realistic apprehensions (from historical experience) that the compounds, which can reduce mutations and cancer rates, ".. would have some easily discovered deleterious side effects. However small this might be, it would almost certainly prevent the use of the substance" (Crow 1986). Thus, all the factors must be elucidated before antimutagens and anticarcinogens can be applied in practice.

No attempt was made here to transfer all the concepts of antimutagenic research, which were mostly developed for gene mutations in microorganisms, to the important, but yet separate field of clastogenesis. However, we believe that in the future the intense research in antimutagenesis will include the estimation of anticlastogenesis. Then "desclastogens" and "bioanticlastogens" will become common terms for the respective agents. Antimutagenesis is an integrative approach toward protection of hereditary structures. We hope that our book will be useful in the complex and difficult process of its understanding.

References

Abe S (1986) Effects of arachidonic acid and indomethacin on sister-chromatid exchange induction by polycyclic aromatic hydrocarbons in mammalian cell lines. Mutation Res 173:55–60

Abel G, Mannschedel A (1987) The clastogenic effect of 5-methoxy-psoralene plus UV-A in human lymphocytes in vitro and its modification by the anticlastogen βaminoethylisothiouronium. Hum Genet 76:181–185

Abka'i F, Wachter E, Tittelbach H, Gebhart E (1987) Induction by chemical clastogens of aberrations in prematurely condensed interphase chromatin of Chinese hamster ovary cells. Cell Biol Toxicol 3:285–301

Abraham SK, Mahajan S, Kesavan PC (1986) Inhibitory effects of dietary vegetables on the in vivo clastogenicity of cyclophosphamide. Mutation Res 172:51–54

Abramovsky I, Vorsanger G, Hirschhorn K (1978) Sister-chromatid exchange induced by X-ray of human lymphocytes and the effect of l-cysteine. Mutation Res 50:93–100

Adler I-D, Ashby J, Würgler FE (1989) Screening for possible human carcinogens and mutagens: a symposium report. Mutation Res 213:27–39

Alekperov U (1984) Antimutagenesis: theoretical and practical aspects. Nauka, Moscow (in Russian)

Ames BN (1983) Dietary carcinogens and anticarcinogens: oxygen radicals and degenerative diseases. Science 221:1256–1264

Ames BN (1984) Cancer and diet. Science 224:668–670; 757–760

Ames BN (1986) Carcinogens and anticarcinogens. In: Shankel DM, Hartman PE, Kada T, Hollaender A (eds) Antimutagenesis and anticarcinogenesis mechanisms. Plenum, New York pp 7–35

Ames BN, Durston WE, Yamasaki E, Lee FD (1973) Carcinogens are mutagens: a simple test system combining liver homogenates for activation and bacteria for detection. Proc Natl Acad Sci USA 70:2281–2285

Amonkar AJ, Padma PR, Bhide SV (1989) Protective effect of hydroxy-chavicol, a phenolic component of betel leaf, against the tobacco-specific carcinogens. Mutation Res 210: 249–253

Anderson D, Richardson CR, Purchase IFH, Evans HJ, O'Riordan ML (1981) Chromosomal analysis in vinyl chloride exposed workers: comparison of the standard technique with the sister-chromatid exchange technique. Mutation Res 83:137–144

Anderson D, Jenkinson PC, Dewdney RS, Francis AJ, Godbert P, Butterworth KR (1988) Chromosome aberrations, mitogen-induced blastogenesis and proliferative rate index in peripheral lymphocytes from 106 control individuals of the U.K. population. Mutation Res 204:407–420

Archer PG, Bender M, Bloom AD, Brewen JG, Carrano AV, Preston RJ (1981) Report of Panel 1: guidelines for cytogenetic studies in mutagen-exposed human populations. In: Bloom AD (ed) Guidelines for studies of human populations exposed to mutagenic and reproductive hazards. March of Dimes Birth Defects Found, New York, pp 1–35

Arlett CF, Lehmann AR (1978) Human disorders showing increased sensitivity to the induction of genetic damage. Ann Rev Genet 12:95–115

Arutyunyan RM (1973) L-cysteine modification of the cytogenetic effect of cytosine-arabinoside in human peripheral blood lymphocyte cultures. Byull Eksp Biol Med 75:103–105 (in Russian)

Arutyunyan RM (1985) Modification of chemical mutagenesis in human cells. Univ Yerevan Press, Yerevan (in Russian)

Arutyunyan RM, Egiazarian SV (1975) Investigation of radioprotectors cytogenetic activity in human lymphocyte culture Tsitol Genet 9/4:295–298 (in Russian)

Arutyunyan RM, Kuleshov NP (1972) Modification by cysteine of tio-TEF induced chromosome aberrations in cultured human leukocytes Soviet Genet 8:534–538

Arutyunyan RM, Oganyan VK (1975) Investigation of the cytogenetic activity of radioprotectors in human lymphocyte cultures. Communication III. Tsitol Genet 9/5:396–399 (in Russian)

Arutyunyan RM, Sarkisian TF (1978) Models of modification of chemical mutagenesis in human cells. II. Correlation of the mutagenic effect with the time of the culture treatment. Genetika 14:2205–2208 (in Russian)

Arutyunyan RM, Zalinyan GG (1987) Modifying action of caffeine on the cytogenetic activity of gibberellic acid in human lymphocytes culture. Biol J Armenia 40:153–155 (in Russian)

Arutyunyan RM, Oganyan VK, Sarkisian TF (1977) Modelling of a modification of chemical mutagenesis in human cells. I. Elimination of chromosome aberrations. Genetika 13/9:1632–1637 (in Russian)

Asano M, McDonald TP, Odell TT (1961) Effects of nitrogen mustard on mouse tissue and modification by AET. Int J Radiat Biol 4:591–600

Ashby J (1988) Comparison of techniques for monitoring human exposure to genotoxic chemicals. Mutation Res 204:543–551

Ashby J, Richardson CR (1985) Tabulation and assessment of 113 surveillance cytogenetic studies conducted between 1965 and 1984. Mutation Res 154:111–133

Au WW, Anwar WA, Hanania E, Ramanujam S (1990) Antimutagenic effects of dimethyl sulfoxide on metabolism and genotoxicity of benzene in vivo. In: Kuroda Y, Shankel DM, Waters MD (eds) Antimutagenesis and anticarcinogenesis mechanisms II. Plenum, New York, pp 389–393

Auf der Maur P, Berlincourt-Böhni K (1979) Human lymphocyte cell cycle: studies with the use of BrUdR. Hum Genet 49:209–215

Avanzi S (1957) Inibizione dell' effetto citologico dell' acetato di Tallio per mezzo della cisteine. Caryologia 10:96–101

Avanzi S (1961) Chromosome breakage by pyrrolizidine alkaloids and modification of the effect with cysteine. Caryologia 14:251–261

Awogi T, Sato T (1989) Micronucleus test with benzo(a)pyrene using a single peroral administration and intraperitoneal injection in males of the Ms/Ae and CD-1 mouse strains. Mutation Res 223:353–356

Bajrakova AK, Pantev TP, Nikolov IT, Bokova N (1974) Protection of mouse spermatogonia against X-ray-induced translocations. Mutation Res 25:377–381

Banduhn N, Obe G (1985) Mutagenicity of methyl 2-benzimidazolecarbamate, diethylstilbestrol and estradiol: structural chromosome aberrations, sister chromatid exchanges, C-mitoses, poliploidies and micronuclei. Mutation Res 156:199–218

Barbacid M (1986) Mutagens, oncogenes and cancer. Trends Genet 2/7:188–192

Baret A, Emerit I (1983) Variation of superoxide dismutase levels in fetal calf serum. Mutation Res 121:293–297

Barthelmeß A (1959) Gefährliche Dosis? Erbgesundheit im technischen Zeitalter. Herder, Freiburg

Bartsch H, Oshima J, Nair J, Pignatelli B, Calmels S (1986) Modifiers of endogenous nitrosamine synthesis and metabolism. In: Shankel DM, Hartman PE, Kada T, Hollaender A (1986) Antimutagenesis and anticarcinogenesis mechanisms. Plenum, New York, pp 87–101

Becher R, Gebhart E (1973) Die Schutzwirkung von Cysteamin und β-Aminoäthylisothiouronium (AET) gegen die chromosomenschädigende Aktivität von Trenimon in menschlichen Lymphocyten in vitro. Humangenetik 17:307–316

Beek B, Obe G (1974) The human leukocyte test system. II. Different sensitivities of sub-populations to a chemical mutagen. Mutation Res 24:395–398

Beek B, Klein G, Obe G (1980) The fate of chromosomal aberrations in a proliferating cell system. Biol Zbl 99:73–84

Bellett AJD, Waldron-Stevens LK, Braithwaite AW, Cheetham BF (1982) Spermine and aminoguanidine protect cells from chromosome aberrations induced by adenovirus during the G2 phase of the cell cycle. Chromosoma 84:571–583

Bender MA, Brewen JG (1969) Factors influencing chromosome aberration yields in the human peripheral leukocyte system. Mutation Res 8:383–399

Bender MA, Griggs HG, Bedford JS (1974) Mechanisms of chromosomal aberration production. III. Chemicals and ionizing radiation. Mutation Res 23:197–212

Bender MA, Preston RJ, Leonard RC, Pyatt BE, Gooch PC (1989) Chromosomal aberration and sister chromatid exchange frequencies in peripheral blood lymphocytes of a large human population sample. Mutation Res 212:149–154

Benova DK (1986) Antimutagenic properties of selected radioprotective drug mixtures with regard to X-ray-induced reciprocal translocations in mouse spermatogonia. Mutation Res 159:75–81

Bertram JS, Kolonel LN, Meyskens FL (1987) Rationale and strategies for chemoprevention of cancer in humans. Cancer Res 47:3012–3031

Bianchi M, Bianchi NO, Brewen JG, Buckton KE, Fabry L, Fischer P, Gooch PC, Kucerova M, Leonard A, Mukherjee RN, Mukherjee U, Nakai S, Natarajan AT, Obe G, Palitti F, Pohl-Rüling J, Schwarzacher HG, Scott D, Sharma T, Takahashi E, Tanzarella C, Van Buul PPW (1982a) Evaluation of radiation-induced chromosomal aberrations in human peripheral blood lymphocytes in vitro. Result of an IAEA-coordinated programme. Mutation Res 96:233–242

Bianchi M, Bianchi N, Cortes L, Reigosa M (1982b) Cysteamine protection of SCEs induced by UV and fluorescent light. Mutation. Res 104:281–286

Bochkov NP, Yakovenko KN, Chebotarev AN, Funes-Cravioto F, Zhurkov VS (1972) Distribution of defective chromosomes in human cells after treatment with chemical mutagens in vitro and in vivo. Soviet Genet 8:1595–1601

Bochkov NP (1982) Medical and biological approaches to cytogenetic monitoring of human populations. In: Sorsa M, Vainio H (eds) Mutagens in our environment. Liss, New York, pp 101–109

Bochkov NP, Chebotarev AN (1989) Human heredity and environmental mutagens. Meditsina, Moscow (in Russian)

Bochkov NP, Yakovenko KN (1978) Sensitivity of human lymphocyte chromosomes to ethylenimine derivatives at different periods during cell culture. In: Evans HJ, Lloyd DC (eds) Mutagen-induced chromosome damage in man. Univ Press, Edinburgh, pp 290–299

Bochkov NP, Yakovenko KN, Nazarenko SA (1979) Combined action of alkylating compounds on human chromosomes. Genetika 15:162–170 (in Russian)

Bochkov NP, Stukalov SV, Chebotarev AN (1982) Comparison of the rate of chromosome aberrations induced by thiophosphamide in rabbit lymphocytes in vivo and in vitro. Bull Exp Biol Med 34:90–92

Bochkov NP, Chebotarev AN, Filippova TV, Platonova VI, Stukalov SV, Debova GA (1984) Alterations in the baseline sister-chromatid exchange frequency in human lymphocyte culture following a number of cell divisions. Mutation Res 127:149–153

Bochkov NP, Filippova TV, Kuzin SM, Stukalov SV (1986) Cytogenetic effects of cyclophosphamide on human lymphocytes in vivo and in vitro. Mutation Res 159:103–107

Boller K, Schmid W (1970) Chemische Mutagenese beim Säuger. Das Knochenmark des Chinesischen Hamsters als in vivo-Testsystem. Hämatologische Befunde nach Behandlung mit Trenimon. Humangenetik 11:35–54

Bonati F, Savino G (1965) Thiol compounds as protectors against X-rays and radiomimetic agents. Prog Biochem Pharmacol 1:489–500

Boone CW, Kelloff GJ, Malone WE (1990) Identification of candidate cancer chemopreventive agents and their evaluation in animal models and human clinical trials: a review. Cancer Res 50:2–9

Braun W, Stille G, Wolf V (1957) Oraler und parenteraler Strahlenschutz durch Thiolaktone. Arzneimittel-Forsch 7:753

Bridges BA (1980) An approach to the assessment of the risk to man from DNA damaging agents. Arch Toxicol 3 Suppl, pp 271–281

Brusick D (1980) Principles of genetic toxicology. Plenum, New York

Buckton KE, Evans HJ (eds) (1973) Methods for the analysis of human chromosome aberrations. WHO, Geneva

Buckton KE, Evans HJ (1982) Human peripheral blood lymphocyte cultures: an in vitro assay for the cytogenetic effects of environmental mutagens. In: Hsu TC (ed) Cytogenetic assays of environmental mutagens. Allanheld, Osmun, Totowa NJ, pp 183–202

Bueding E, Ansher S, Dolan P (1986) Anticarcinogenic and other protective effects of dithiolthiones. In: Shankel DM, Hartman PE, Kada T, Hollaender A (eds) Antimutagenesis and anticarcinogenesis mechanisms. Plenum, New York, pp 483–489

Carrano AV (1982) Sister chromatid exchange as an indicator of human exposure. In: Bridges BA, Butterworth BE, Weinstein IB (eds) Indicators of genotoxic exposure. Banbury Report 13, Cold Spring Harbor, pp 307–316

Carrano AV, Natarajan AT (1988) Considerations for population monitoring using cytogenetic techniques. ICPEMC Publ No 14. Mutation Res 204:379–406

Carrano AV, Thompson LH (1982) Sister chromatid exchange and gene mutation. Cytogenet Cell Genet 33:57–61

Chang TH, Klassen W (1968) Comparative effects of tretamine, tepa, apholate, and their structural analogs on human chromosomes in vitro. Chromosoma 24:314–323

Chaudhuri JP, Langendorff H (1968) Chemical radioprotection of mammalian chromosomes in vivo: radioprotection of rat bone-marrow chromosomes with a single prophylactic dose of AET. Int J Radiat Biol 14:463–467

Chebotarev AN (1985) Quantitative rules of formation of sister chromatid exchanges. Inst Med Genet Acad Med Sci USSR Moscow, p 46 (in Russian)

Chebotarev AN, Yakovenko KN (1974) Simulation of the dependence of cytogenetic effects on the mutagenic concentration. Soviet Genet 10:1048–1054

Chen JF, Lin YJ (1987) Effects of amino acid-supplemented diets on sister chromatid exchange frequency of rats treated with cyclophospamide. Cytologia 52:687–692

Chen LH, Boissonneault GA, Glauer HP (1988) Vitamin C, vitamin E and cancer (review). Anticancer Res 8:739–748

Clarke CH, Shankel DM (1975) Antimutagenesis in microbial systems. Bacteriol Rev 39:33–53

Clarkson JM, Humphrey RM (1976) The significance of DNA-damage in the cell cycle sensitivity of Chinese hamster ovary cells to bleomycin. Cancer Res 36:2345–2349

Cleaver JE, Arutyunyan RM, Sarkisian T, Kaufmann WK, Greene AE, Coriell L (1980) Similar defects in DNA repair and replication in the pigmented xerodermoid and the xeroderma pigmentosum variants. Carcinogenesis 1:647–655

Collaborative Study Group for the Micronucleus Test (1986) Sex difference in the micronucleus test. Mutation Res 172:151–163

Collaborative Study Group for the Micronucleus Test (1988) Strain difference in the micronucleus test. Mutation Res 204:307–316

Countryman PI, Heddle JA (1976) The production of micronuclei from chromosome aberrations in irratiated cultures of human lymphocytes. Mutation Res 41:321–332

Cozzi R, Bona R, Cundari E, Perticone P (1989) Influence of spermidine on sister chromatid exchanges induced by alkylating agents in mammalian cells in vitro. Anticancer Res 9:1129–1132

Crow JF (1986) Population consequences of mutagenesis and antimutagenesis. In: Shankel DM, Hartman PE, Kada T, Hollaender A (eds) Antimutagenesis and anticarcinogenesis mechanisms. Plenum, New York, pp 519–530

Dallapiccola B, Porfirio B, Mokini V, Alimena G, Isacchi G, Gandini E (1985) Effects of oxidants and antioxidants on chromosomal breakage in Fanconi anemia lymphocytes. Hum Genet 69:62–65

Danielian EA, Arutyunyan RM, Ispiryan SM (1978) Study of the cytogenetic activity of radioprotectors in a human lymphocyte culture. Communication VI. Cytol Genet 12:51–55

Darroudi F, Targa H, Natarajan AT (1988) Influence of dietary carrot on cytostatic drug activity of cyclophosphamide and its main directly acting metabolite: induction of sister chromatid exchanges in normal human lymphocytes, Chinese hamster ovary cells, and their DNA repair-deficient cell lines. Mutation Res 198:327–335

De Flora S (1988) Problems and prospects in antimutagenesis and anticarcinogenesis (editorial). Mutation Res 202:279–283

De Flora S, Ramel C (1988) Mechanisms of inhibitors of mutagenesis and carcinogenesis. Classification and overview. Mutation Res 202:285–306

De Jong G, Van Sittert NJ, Natarajan AT (1988) Cytogenetic monitoring of industrial populations potentially exposed to genetoxic chemicals and of control populations. Mutation Res 204:451–464

Devik F (1954) Protective effects of combined hypoxia and cysteine treatment on whole-body irradiation of mice. Brit J Radiol N S 27:463–466

Devik F, Lothe F (1955) The effect of cysteamine, cystamine, and hypoxia on mortality and bone marrow chromosome aberrations in mice after total roentgen irradiation. Acta Radiol 44:243–248

Ehling UH (1964a) AET-Schutzwirkung bei strahleninduzierten dominanten Letalmutationen der Maus. Strahlentherapie 125:128–135

Ehling UH (1964b) Schutzwirkung von Cystein und AET bei strahleninduzierten dominanten Letalmutationen der Maus. Biophysik 1:427–428

Ehling UH (1976) Mutagenicity testing and risk estimation with mammals. Mutation Res 41:113–122

Ehling UH (1988) Quantification on the genetic risk of environmental mutagens. Risk Analysis 8:45–57

Eidus LKl (1977) Nonspecific reaction of cells and radiosensitivity. Atomizdat, Moscow 1977 (in Russian)

Eldjarn L, Jellum E (1969) Studies on biochemical and radioprotective effects of thiols and disulphides at the subcellular and molecular level. In: IAEA: radiation damage and sulphydryl compounds. IAEA, Vienna, pp 45–54

Emerit I (1984) Active oxigen species at the origin of sister chromatid exchanges. In: Tice RR, Hollaender A (eds) Sister chromatid exchanges: Part A. Plenum, New York, pp 127–140

Emerit I, Cerutti P (1981) Clastogenic activity from Bloom syndrome fibroblast cultures. Proc Natl Acad Sci USA 78:1868–1872

Emerit I, Levy A, Housset E (1974) Breakage factor in systemic sclerosis and protector effect of l-cysteine. Humangenetik 25:221–226

Emerit I, Emerit J, Levy A, Keck M (1979) Chromosomal breakage in Crohn's disease: anticlastogenic effect of d-penicillamine and l-cysteine. Hum Genet 50:51–57

Emerit I, Michelson AM, Levy A, Camus JP, Emerit J (1980) Chromosome breaking agent of low molecular weight in human systemic lupus erythematosus. Hum Genet 55:341–344

Emerit I, Levy A, Michelson AM (1981) Effect of superoxide dismutase on the chromosomal instability of New Zealand black mice. Cytogenet Cell Genet 30:65–69

Emerit I, Levy A, Cerutti P (1983) Suppression of tumor promoter phorbolmyristate acetate-induced chromosome breakage by antioxidants and inhibitors of arachidonic acid metabolism. Mutation Res 110:327–335

Evans HJ (1982a) Cytogenetic studies on industrial populations exposed to mutagens. In: Bridges BA, Butterworth BE, Weinstein IB (eds) Indicators of genotoxic exposure. Banbury Rep No 13. Cold Spring Harbor, pp 325–336

Evans HJ (1982b) Chromosomal mutations in human populations. Cytogenet Cell Genet 33:48–56

Evans HJ, O'Riordan ML (1975) Human peripheral blood lymphocytes for the analysis of chromosome aberrations in mutagen tests. Mutation Res 31:149–152

Farber E (1984) The multistep nature of cancer development. Cancer Res 44:4217–4223

Fearon ER, Vogelstein B (1990) A genetic model for colorectal tumorigenesis. Cell 61:759–767

Fenech M, Morley AA (1985) Cytokinesis-block micronucleus method in human lymphocytes: effect of in vivo ageing and low dose X-irradiation. Mutation Res 161:193–198

Firket H, Mahieu P (1966) Irradiation et protection de cultures synchrones de cellules HeLa; effet sur le premier cycle cellulaire. Int J Radiat Biol 11:245

Forni A (1984) Chromosomal aberrations in monitoring exposure to mutagens-carcinogens. In: Berlin A, Draper M, Hemminki K, Vainio H (eds) Monitoring human exposure to carcinogenic and mutagenic agents. IARC Sci Publ No 59:325–337

Forni A, Bertazzi PA (1987) Epidemiology of protection and prevention against environmental mutagens/carcinogens. Examples from occupational medicine. Mutation Res 181:289–297

Forni AM, Capellini A, Pacifico E, Vigliani EC (1971) Chromosome changes and their evolution in subjects with past exposure to benzene. Arch Environ Health 23:385–391

Frank DW, Trzos RJ, Good PI (1978) A comparison of two methods for evaluating drug-induced chromosome alterations. Mutation Res 56:311–317

Gad-El-Karim MM, Harper BL, Legator MS (1984) Modifications of the myeloclastogenic effect of benzene in mice with toluene, phenobarbital, 3-methylcholanthrene, Aroclor 1254 and SKF-525A. Mutation Res 135:225–243

Galdean D, Petrasincu D, Alangiu P, Ibric S, Voiculetz N (1986) The protective effect of Thiola against the genotoxic action of benzo<a>pyrene. Experientia 42:572–574

Galloway SM, Painter RB (1979) Vitamin C is positive in the DNA synthesis inhibition and sister chromatid exchange test. Mutation Res 60:321–327

Galloway SM, Bloom AD, Resnick M, Margolin BH, Nakamura F, Archer P, Zeiger E (1985) Development of a standard protocol for in vitro cytogenetic testing with Chinese hamster ovary cells: comparison of results for 22 compounds in two laboratories. Environ Mutagen 7:1–51

Galloway SM, Armstrong MJ, Reuben C, Colman S, Cannon C, Bloom AD, Nakamura F, Ahmed M, Duk S, Rimpo J, Margolin BH, Resnick MA, Anderson B, Zeiger E (1987) Chromosome aberrations and sister chromatid exchanges in Chinese hamster ovary cells: evaluation of 108 chemicals. Environ Molec Mutagen 10 Suppl, pp 1–175

Garner RC (1980) In vitro assays to predict carcinogenicity? In: Norpoth KH, Garner RC (eds) Short-term test systems for detecting carcinogens. Springer, Berlin Heidelberg New York, pp 5–18

Garriott ML, Crowe DT (1983) AET reduces the frequency of micronuclei in bone marrow cells of mice exposed to gamma radiation. Radiat Res 93:200–204

Gebhart E (1968) Die Wirkung von l-Cystein auf die chromatidbrechende Aktivtät von 8-Hydroxychinolinsulfat. Protoplasma 65:247–250

Gebhart E (1969) Zur Beeinflussung der Wirkung von Myleran auf menschliche Chromosomen durch L-Cystein. Mutation Res 7:254–257

Gebhart E (1970) The treatment of human chromosomes in vitro: results. In: Vogel F, Röhrborn G (eds) Chemical mutagenesis in mammals and man. Springer, Berlin Heidelberg New York, pp 367–382

Gebhart E (1971) Die Wirkung von L-Cystein auf die Aberrationsauslösung in menschlichen Chromosomen durch chemische Mutagene. II. Zur Bedeutung des Zeitfaktors bei Verwendung von Trenimon als Aberrationsinduktor. Humangenetik 11:237–243

Gebhart E (1972) Untersuchungen über die Schutzwirkung einiger Radioprotektoren und Aminosäuren gegen die chromosomenschädigende Aktivität chemischer Mutagene in menschlichen Lymphocyten in vitro. Habilitationsschrift, Erlangen

Gebhart E (1973a) Untersuchungen über die Beeinflussung der chromosomenschädigenden Aktivität von Trenimon an menschlichen Lymphocyten in vitro durch Aminosäuren. Humangenetik 18:237–246

Gebhart E (1973b) Vergleichende Untersuchungen über die Beeinflussung der chromosomenschädigenden Aktivität von 8-Hydroxychinolinsulfat in menschlichen Lymphocyten in vitro durch Radioprotektoren und Aminosäuren. Mutation Res 18:353–361

Gebhart E (1974) Antimutagens. Data and problems. Humangenetik 24:1–32

Gebhart E (1977) Chemische Mutagenese. Fischer, Stuttgart

Gebhart E (1978) The anticlastogenic effect of various combinations of cysteamine, AET, HCT, and amino acids on chromosome damage by trenimon and bleomycin in human lymphocytes in vitro. Hum Genet 43:185–203

Gebhart E (1981) Sister chromatid exchange (SCE) and structural chromosome aberration in mutagenicity testing. Hum Genet 58:235–254

Gebhart E (1982) The epidemiological approach: chromosome aberrations in persons exposed to chemical mutagens. In: Hsu TC (ed) Cytogenetic assays of environmental mutagens. Allanheld, Osmun Totowa, NJ, pp 385–408

Gebhart E (1984) The action of anticlastogens on chemically induced SCE. In: Tice RR, Hollaender A (eds) Sister chromatid exchanges. Pt A. Plenum, New York, pp 319–332

Gebhart E (1989) Genetic alterations in cancer cells -- what do they tell us for mutagenicity testing? Toxicol Environ Chem 19:7–15

Gebhart E (1990a) Cytogenetics and molecular genetics of cell transformation in vitro by mutagens/carcinogens. In: Obe G (ed) Advances in mutagenesis research, vol 2. Springer, Berlin Heidelberg New York, pp 275–302

Gebhart E (1990b) Tumorzytogenetik. Schattauer, Stuttgart

Gebhart E, Becher R (1976) Comparative studies on the specificity of anticlastogenic action in human lymphocytes in culture. Hum Genet 32:49–64

Gebhart E, Kappauf H (1980) The action of three anticlastogens on the induction of sister chromatid exchange by trenimon and 8-hydroxyquinoline sulfate in human lymphocyte cultures. Environ Mutagen 2:191–200

Gebhart E, Windolph B, Wopfner F (1980) Chromosome studies in lymphocytes of patients under cytostatic therapy. II. Studies using BUDR-labelling technique in cytostatic interval therapy. Hum Genet 56:157–167

Gebhart E, Wagner H, Behnsen H (1984) The action of anticlastogens in human lymphocyte cultures and its modification by rat liver S9 mix. I. Studies with AET and sodium fluoride. Mutation Res 129:195–206

Gebhart E, Wagner H, Grziwok K, Behnsen H (1985) The action of anticlastogens in human lymphocyte cultures and their modification with rat liver S9 mix. II. Studies with vitamins C and E. Mutation Res 149:83–94

German J (1969) Chromosomal breakage syndromes. Birth Def Orig Art Ser 5:117–131

Ghaskadbi S, Vaidya VG (1989) In vivo antimutagenic effect of ascorbic acid against mutagenicity of the common antiamebic drug diiodohydroxyquinoline. Mutation Res 222:219–222

Ghaskadhi S, Pavaskar SV, Vaidya VG (1987) Bio-antimutagenic effect of l-cysteine on diiodohydroxyquinoline-induced micronuclei in Swiss mice. Mutation Res 187:219–222

Gilbert MM, Sprecher J, Chang LW, Meisner LF (1983) Protective effect of vitamin E on genotoxicity of methylmercury. J Toxicol Environ Health 12:767–773

Glatt HR, Oesch F (1986) Structural and metabolic parameters governing the mutagenicity of polycyclic hydrocarbons. In: De Serres FJ (ed) Chemical Mutagens, Vol 10, Plenum, New York, 73–127

Grady MK, Jacobson-Kram D, Dearfield KL, Williams JR (1986) Induction of sister chromatid exchanges by benzidine in rat and human hepatoma cell lines and inhibition by indomethacin. Cell Biol Toxicol 2:223–230

Gugushvili BS, Pomerantzeva MD, Vilkina GA (1972) The protective effect of cystamine on X-ray-induced damages in germ cells of male mice. Genetika 8:42–49 (in Russian)

Gurtoo HL, Marinello AJ, Berigan MJ, Bansal SK, Paul B, Pavelic ZP, Struck RF (1983) Effect of thiols on toxicity and carcinostatic activity of cyclophosphamide. Sem Oncol 10, Suppl 1:35–45

Gustavino B, Johannes C, Obe G (1986) Restriction endonuclease Bam HI induces chromosomal aberrations in Chinese hamster ovary (CHO) cells. Mutation Res 175:91–95

Hadnagy W, Seemayer NH (1988) Antimutagenicity of chlorophyllin against airborne pollutants. Mutation Res 203:205–206

Hamilton HB (1982) Genetic markers in the atomic bomb survivors and their children — Hiroshima and Nagasaki. Jpn J Hum Genet 27:113–119

Hanawalt PC (1989) Concepts and models for DNA repair: *Escherichia coli* to mammalian cells. Environ Mol Mutagen 14 Suppl 16:90–98

Hanawalt PC, Sarasin A (1986) Cancer-prone hereditary diseases with DNA processing abnormalities. Trends Genet 5:124–129

Hanck A (1985) Ascorbic acid and cancer. In: Meyskens FL, Prasad KN (eds) Vitamins and cancer — human cancer prevention by vitamins and micronutrients. Humana, Clifton, pp 365–397

Hansteen IL (1982) SCE as a monitor of industrial and environmental toxins. In: Sandberg AA (ed) Sister chromatid exchange. Liss, New York, pp 675–697

Harnden DG (1976) Chromosome abnormalities and pre-disposition towards cancer. Proc Soc Med 69:41–43

Harper BL, Legator MS (1987) Pyridine prevents the clastogenicity of benzene but not of benzo<a>pyrene or cyclophosphamide. Mutation Res 179:23–31

Harrison EF, Fuquay ME, Hunter HL (1983) Effect of N-acetylcysteine on the antitumor activity of cyclophosphamide against Walker-256 carcinosarcoma in rats. Sem Oncol 10, Suppl 1:25–28

Hartman PE, Shankel DM (1990) Antimutagens and anticarcinogens: a survey of putative interceptor molecules. Environ Molec Mutagen. 15:145–182

Hasegawa J, Hosokawa M, Okada F, Kobayashi H (1989) Inhibition of mitomycin C-induced sister-chromatid exchanges in mouse bone marrow cells by the immunopotentiators Krestin and Lentinan. Mutation Res 226:9–12

Hayashi M, Sutou S, Shimada H, Sato S, Sasaki YF, Wakata A (1989) Difference between intraperitoneal and oral gavage application in the micronucleus test. Mutation Res 223:329–344

Hayatsu H, Arimoto S, Negishi T (1988) Dietary inhibitors of mutagenesis and carcinogenesis. Mutation Res 202:429–446

Heddle JA, Harris JW (1975) Rapid screening of radioprotective drugs in vivo. Radiat Res 61:350–353

Heddle JA, Krepinsky AB, Marshall RR (1983) Cellular sensitivity to mutagens and carcinogens in chromosome-breakage and other cancer-prone syndromes. In: German J (ed) Chromosome mutation and neoplasia. Liss, New York, pp 203–239

Heddle JA, Stuart E, Salamone MF (1984) The bone marrow micronucleus test. In: Kilbey BJ, Legator M, Nichols W, Ramel C (eds) Handbook of mutagenicity test procedures. Elsevier, Amsterdam, pp 441–457

Heim S, Mitelman F (1987) Cancer cytogenetics. Liss, New York

Heindorff K, Rieger R, Schubert I, Michaelis A, Aurich O (1987) Clastogenic adaptation of plant cells – reduction of the yield of clastogen-induced chromatid aberrations by various pretreatment procedures. Mutation Res 181:157–171

Hittelman WN, Pollard M (1982) A comparison of the DNA and chromosome repair kinetics after γ-irradiation. Radiat Res 92:497–509

Hittelman WN, Rao PN (1974) Premature chromosome condensation. I. Visualization of X-ray induced chromosome damage in interphase cells. Mutation Res 23:251–258

Hittelman WN, Rao PN (1975) The nature of adriamycin-induced cytotoxicity in Chinese hamster cells as revealed by premature chromosome condensation. Cancer Res 35:3027–3035

Hochstein P, Atallah AS (1988) The nature of oxidants and antioxidants systems in the inhibition of mutation and cancer. Mutation Res 202:363–375

Högstedt B, Gullberg B, Mark-Vendel E, Mitelman F, Skerfving S (1981) Micronuclei and chromosome aberrations in bone marrow cells and lymphocytes of humans exposed mainly to petroleum vapors. Hereditas 94:179–187

Högstedt B, Gullberg B, Hedner K, Kolnig AM, Mitelman F, Skerfving S, Widegren B (1983) Chromosome aberrations and micronuclei in bone marrow cells and peripheral blood lymphocytes in humans exposed to ethylene oxide. Hereditas 98:105–113

Hollaender A (ed) (1971) Chemical mutagens. Principles and methods of their detection, vols 1, 2. Plenum, New York

Holmes GE, Holmes FF (1978) Pregnancy outcome of patients treated for Hodgkin's disease. Cancer 41:1317–1322

Hsu TC (ed) (1982) Cytogenetic assays of environmental mutagens. Allanheld, Osmun, Totowa, NJ

Hsueh JL, Chen HH, Huang CC (1981) Effect of tumor promoter 12-o-tetradecanoyl-phorbol-13-acetate on induction of sister chromatid exchanges in Chinese hamster V79 cells treated with mutagens. Mutation Res 81:387–394

Hu G (1985) Influence of sodium selenite on the number of sister chromatid exchanges in V79 cells induced by MNNG and MNU. Zhonghua Zhongliu Zazhi 7:22–25

Huang CC (1987) Retinol (vitamin A) inhibition of dimethylnitrosamine (DMN) and diethylnitrosamine (DEN) induced sister-chromatid exchanges in V79 cells and mutations in *Salmonella*/microsome assays. Mutation Res 187:133–140

Huang CC, Hsueh JL, Chen HH, Batt TR (1982) Retinol (vitamin A) inhibits sister chromatid exchanges and cell cycle delay induced by cyclophosphymide and aflatoxin B1 in Chinese hamster V79 cells. Carcinogenesis 3:1–5

Huber R, Bauchinger M (1990) Development and perspectives of the human lymphocyte micronucleus assay. In: Obe G (ed) Advances in mutagenesis research. Springer, Berlin Heidelberg New York, pp 89–104

Hulten M, Lindsten J, Lidberg L, Ekelund H (1968) Studies on mitotic and meiotic chromosomes in subjects exposed to LSD. Ann Genet 11:201–210

Inoue K, Shibata T, Kosaka H, Uozumi M, Tsuda S, Abe T (1985) Induction of sister chromatid exchanges by N-nitrosocimetidine in cultured human lymphocytes and its inhibition by chemical compounds. Mutation Res 156:117–121

Inouye T, Sasaki YF, Imanishi H, Watanabe M, Ohta T, Shirasu Y (1988) Suppression of mitomycin C-induced micronuclei in mouse bone marrow cells by post-treatment with vanillin. Mutation Res 202:93–95

Ishii Y, Bender MA (1980) Effects of inhibitors of DNA synthesis on spontaneous and ultraviolet light-induced sister-chromatid exchanges in Chinese hamster cells. Mutation Res 79:19–32

Iskandar O (1979) An improved method for the detection of micronuclei in human lymphocytes. Stain Technol 54:221–223

Ito N, Fukushima S, Tsuda H (1985) Carcinogenicity and modification of the carcinogenic response by BHA, BHT, and other antioxidants. CRC Crit Rec Toxicol 15:109–150

Ito Y, Maeda S, Sugiyama T (1986) Suppression of 7,12-dimethylbenz <a>anthracene-induced chromosome aberrations in rat bone marrow by vegetable juices. Mutation Res 172:55–60

Ivett JL, Tice RR (1982) Average generation time: a new method of analysis and quantitation of cellular proliferation kinetics. Environ Mutagen 4:358

Iwawaki H, Ito M, Kogiso S, Yoshitake A (1988) Effects of l-ascorbic acid on the clastogenicity of catecholamines (Abstr). Mutation Res 203:374

Jacky PB, Beek B, Sutherland GR (1983) Fragile sites in chromosomes: possible model for the study of spontaneous chromosome breakage. Science 220:69–70

Jellum E (1965) Interaction of cystamine and cystamine derivatives with nucleic acids and nucleoproteins. Int J Radiat Biol 9:185–200

Jellum E (1966) The role of cystamine-nucleic acid interactions in protection against X ray induced damage of DNA. Int J Radiat Biol 10:577–594

Jellum E, Eldjarn L (1965) Isolation of SH-containing DNA-nucleoproteins by chromatography on organomercurial polysaccharide. Biochim Biophys Acta 100:144

Jenderny J, Röhrborn G (1987) Chromosome analysis of human sperm. I. First results with a modified method. Hum Genet 76:385–388

Jensen MK, Nyfors A (1979) Cytogenetic effect of methotrexate on human cells in vivo. Comparison between results obtained by chromosome studies on bone-marrow cells and blood lymphocytes and by the micronucleus test. Mutation Res 64:339–343

Johnson HG, Bach MK (1967) On the mechanism of antimutagen action. 5th Int Congr Chemother, Vienna

Johnson RT, Rao PN (1970) Mammalian cell fusion: induction of premature chromosome condensation in interphase nuclei. Nature (Lond) 226:717–722

Johnson RT, Collins ARS, Waldren CA (1982) Premature chromosome condensation and the analysis of DNA and chromosome lesions. In: Rao PN, Johnson RT, Sperling K (eds) Premature chromosome condensation. Academic Press, New York, pp 253–303

Kada T, Inoue T, Namiki N (1982) Environmental desmutagens and antimutagens. In: Klekowski EJ (ed) Environmental mutagenesis and plant biology. Praeger, New York, pp 137–151

Kagramanyan MS, Arutyunyan RM, Egiazaryan SV (1979) Modes of modification of chemical mutagenesis in human cells. III. Linear index of protection as a unified criterion of modification. Soviet Genet 15:56–63

Kahl R (1986) The dual role of antioxidants in the modification of chemical carcinogenesis. Environ Sci Health C4(1):47–92

Kalweit S, Vasudev V, Obe G (1988) Liquid-holding experiments with human lymphocytes. III. Experiments with G0 and G1 cells. Mutation Res 207:41–44

Kaplan WD, Lyon MF (1953) Failure of mercaptoethylamine to protect against the mutagenic effects of radiation. II. Experiments with mice. Science 118:777–778

Kartzevski E, Gebhart E (1987) Investigations on the anticlastogenic action of N-acetylcysteine in human lymphocytes in vitro. Mutation Res 182:284 (Abstr)

Kaul BL (1969) Modification of the radiomimetic effects of nitroso guanidine by cysteine in plants. Experientia 25:74

Kennedy AR, Radner BS, Nagasawa H (1984) Protease inhibitors reduce the frequecy of spontaneous chromosome abnormalities in cells from patients with Bloom's syndrome. Proc Natl Acad Sci USA 81:1827–1830

Ketterer B (1988) Protective role of glutathione and glutathione transferases in mutagenesis and carcinogenesis. Mutation Res 202:343:361

Kihlman BA (1966) Action of chemicals on dividing cells. Prentice-Hall, Englewood Cliffs

Kihlman BA (1977) Caffeine and chromosomes. Elsevier, Amsterdam

Kilbey BJ, Legator M, Nichols W, Ramel C (eds) (1984) Handbook of mutagenicity test procedures, 2nd edn. Elsevier, Amsterdam

Kinsella AR, Radman M (1980) Inhibition of carcinogen-induced chromosomal aberration by an anticarcinogenic protease inhibitor. Proc Natl Acad Sci USA 77:3544–3547

Kirkland DJ (1989) Mammalian cells in culture. Report of a participant workshop. Mutation Res 213:41–42

Kirkova ZS, Shvetsova TP, Zasukhina GD (1980) Suppressive effect of interferon on the occurrence of chromosome aberrations. Hereditas 93:165–168

Kishi K (1988) Suppressive effect of novobiocin on the frequency of chromosome type aberrations induced by araC in the G1 phase of human lymphocytes. Mutation Res 208:109–113

Kline I, Gang M, Woodman RJ, Cysyk RL, Venditti JM (1973) Protection with N-acetyl-L-cysteine (NSC-111180) against isophosphamide (NSC-109724) toxicity and enhancement of therapeutic effect in early murine L 1210 leukemia. Cancer Chemother Rep Pt 1, 57:299–304

Kodytkova I, Madar J, Sram RJ (1980) Chromosomal aberrations in mouse bone marrow cell, and antibody production changes induced by long-term exposure to cyclophosphamide and a-tocopharol. Folia Biol 26:94–102

Kollmann G, Shapiro B, Martin D (1967) Mechanism of the protective action of GED against radiation damage to DNA. Radiat Res 31:721–731

Korte A, Slacik-Erben R, Obe G (1979) The influence of ethanol treatment on cytogenetic effects in bone marrow cells of Chinese hamsters by cyclophosphamide, Aflatoxin B1, and patulin. Toxicology 12:53–61

Krepinsky AB, Heddle JA (1983) Micronuclei as a rapid and inexpensive measure of radiation-induced chromosomal aberrations. In: Ishihara T, Sasaki MS (eds) Radiation-induced chromosome damage in man. Liss, New York, pp 93–109

Krishna G, Nath J, Ong T (1986) Inhibition of cyclophosphamide and mitomycin C-induced sister chromatid exchanges in mice by vitamin C. Cancer Res 46:2670–2674

Krishna G, Kropko ML, Theiss JC (1989) Use of the cytokinesis-block method for the analysis of micronuclei in V79 Chinese hamster lung cells: results with mitomycin C and cyclophosphamide. Mutation Res 222:63–69

Kukushkina LM, Mosse IB (1975) Comparative study of the effects of the radioprotectant F-37 on the genetic effects of X-rays and the chemical mutagen ThioTEF. Vestsi Akad Navuk BSSR, Ser Biyalag Navuk 2:72–75 (in Russian)

Kuroda Y, Shankel DM, Waters MD (1990) Antimutagenesis and anticarcinogenesis mechanisms II. Plenum, New York

Lähdetie J (1986) Micronucleated spermatids in the seminal fluid of smokers and nonsmokers. Mutation Res 172:255–263

Lähdetie J, Parvinen M (1981) Meiotic micronuclei induced by X-rays in early spermatids of the rat. Mutation Res 81:103–115

Lambert B, Sten M, Hellgren D (1984) Removal and persistence of SCE-inducing damage in human lymphocytes in vitro. In: Tice RR, Hollaender A (eds) Sister chromatid exchanges: Part B. Plenum, New York, pp 647–662

Lamberti L, Bigatti Ponzetto P, Ardito G (1983) Cell kinetics and sister-chromatid-exchange frequency in human lymphocytes. Mutation Res 120:193–199

Larripa I, Mudry de Pargament M, Labal de Vinuesa M, Demattei A, Brieux de Salum S (1984) In vivo and in vitro cytogenetic effects of the anti-tumor agent amsacrina (AMSA). Mutation Res 138:87–91

Latt SA (1973) Microfluorometric detection of deoxyribonucleic acid replication in human metaphase chromosomes. Proc Natl Acad Sci USA 70:3395–3399

Latt SA (1974) Sister chromatid exchanges, indices of human chromosome damage and repair: detection by fluorescence and induction by mitomycin C. Proc Natl Acad Sci USA 71:3162–3166

Lazutka JR, Lekevicius R, Dedonyte V (1988) Sister chromatid exchanges as a test for genotoxicity. Univ Press Vilnius, 77 pp (in Russian)

Le Beau MM (1986) Chromosomal fragile sites and cancer-specific rearrangements. Blood 67:849–858

Lenz W (1979) Medizinische Genetik. 4th ed. Thieme, Stuttgart

Leonard A, Deknudt G (1972) Effects of AET on chromosome rearrangements induced by X-irradiation in spermatogonia. Radiat Res 50:120–124

Leonard A, Deknudt G (1973) Chemical radioprotection against chromosome rearrangements induced in spermatogonia of mice. Strahlentherapie 145:174–177

Leonard A, Maisin JR (1963) Effects of AET against the dominant lethality induced by X-rays in mice. Proc XI Int Congr Genet I, The Hague, p 82

Li FP, Fine W, Jaffe N, Holmes GE, Holmes FF (1979) Offspring of patients treated for cancer in childhood. J Natl Cancer Inst 62:1193–1197

Lialiaris T, Mourelatos D, Dozi-Vassiliades J (1987) Enhancement and attenuation of cytogenetic damage by vitamin C in cultured human lymphocytes exposed to Thiotepa or l-ethionine. Cytogenet Cell Genet 44:209–214

Lindblad A, Lambert B (1981) Relation between sister chromatid exchange, cell proliferation and proportion of B and T cells in human lymphocyte cultures. Hum Genet 57:31–34

Littlefield LG, Colyer SP, Sayer AM, Du Frain RJ (1984) Persistence of SCE-inducing lesions after Go exposure of human lymphocytes to differing classes of DNA-damaging chemicals. In: Tice RR, Hollaender A (eds) Sister chromatid exchanges: Part B. Plenum, New York, pp 663–676

Lüning KG, Frölen H, Nelson A (1961) The protective effect of cysteamine against genetic damages by X-rays in spermatozoa from mice. Radiat Res 14:813–818

Lyn-Cook BD, Patterson RM (1984) Inhibition of free radical induced chromosomal aberrations in vitro by alpha-tocopherol. Nutr Res 4:989–993

Lyon MF (1982) Problems of extrapolation from experimental data to human mutagenesis. In: Sorsa M, Vainio H (eds) Mutagens in our environment. Liss, New York, pp 127–136

Mac Gregor JT, Heddle JA, Hite M, Margolin BH, Ramel C, Salamone MF, Tice RR, Wild D (1987) Guidelines for the conduct of micronucleus assays in mammalian bone marrow erythrocytes. Mutation Res 189:103–112

Mac Rae WD, Stich HF (1979) Induction of sister chromatid exchanges in Chinese hamster ovary cells by the reducing agents bisulfite and ascorbic acid. Toxicology 13:167–174

Makedonov GP, Yakubovskaya EL, Zasukhina GD (1989) Interferon protects human cells against 8-methoxy-psoralen-induced interstrand cross-links by postreplicative repair. Genetika 25:552–544 (in Russian)

Manson MM, Green JA, Driver HE (1987) Ethoxyquin alone induces preneoplastic changes in rat kidney whilst preventing induction of such lesions in liver by aflatoxin B1. Carcinogenesis 8:723–728

Margolin BH, Shelby MD (1985) Sister chromatid exchanges: a reexamination of the evidence for sex and race differences in humans. Environ Mutagen 7 Suppl 4:63–72

Martin RH (1983) A detailed methodology for obtaining preparations of human sperm chromosomes. Cytogenet Cell Genet 35:459–468

Matter B, Schmid W (1971) Trenimon-induced chromosomal damage in bone marrow cells of six mammalian species, evaluated by the micronucleus test. Mutation Res 12:417–425

Mau J (1982) Several approaches to the statistical analysis of data from the sister-chromatid exchange test. In: Müller D, Natarajan AT, Obe G, Röhrborn G (eds) Sister-chromatid-exchange test. Thieme, Stuttgart, pp 79–103

Mc Keown-Eyssen G, Holloway C, Jasmaji V, Brightsee E, Dion P, Bruce WR (1988) A randomized trial of vitamins C and E in the prevention of recurrence of colorectal polyps. Cancer Res 48:4701–4705

Mc Kusick VA (1988) Mendelian inheritance in man, 8th edn. Johns Hopkins Univ Press, Baltimore

Mikaelsen K (1952) The protective effect of glutathione against radiation induced chromosome aberrations. Science 116:172–174

Mikaelsen K (1954) Protective properties of cysteine, sodium hyposulfite, and sodium cyanide against radiation induced chromosome aberrations. Proc Natl Acad Sci USA 40:171–182

Milam KM, Hunter KJ, Deen DF, Marton LJ (1989) Reduction of cisdiamminedichloroplatinum(II)-induced cytotoxicity, sister chromatid exchange and DNA interstrand cross-links in 9L cells treated with the polyamine biosynthesis inhibitor (2R,5R)-6-heptyne-2,5-diamine. Cancer Res 49:6945–6948

Miller EC (1978) Some current perspectives on chemical carcinogenesis in humans and experimental animals: presidential address. Cancer Res 38:1479–1496

Miller LF, Rumack BH (1983) Clinical safety of high oral doses of acetylcysteine. Sem Oncol 10:76–85

Milvy P, Pullman I (1968) ESR studies on spin transfer in irradiated nucleic acid -- cysteamine systems. Radiat Res 34:265

Mitelman F (1985, 1988) Catalog of chromosome aberrations in cancer, 2nd and 3rd edn. Liss, New York

Modig HG, Edgren M, Revesz L (1971) Release of thiols from cellular mixed disulphides and its possible role in radiation protection. Int J Radiat Biol 22:257–268

Morimoto K (1983) Induction of sister chromatid exchanges and cell division delays in human lymphocytes by microsomal activation of benzene. Cancer Res 43:1330–1334

Morimoto K (1984) Proliferative kinetics and chemical-induced sister chromatid exchanges in human lymphocyte cultures. In: Tice RR, Hollaender A (eds) Sister chromatid exchanges: Part B. Plenum, New York, pp 677–693

Morimoto K, Iijima S, Koizumi A (1982) Selenite prevents the induction of sister-chromatid exchanges by methyl mercury and mercuric chloride in human whole-blood cultures. Mutation Res 102:183–192

Morimoto K, Kaneko T, Iijima K, Koizumi A (1984) Proliferative kinetics and chromosome damage in trisomy 21 lymphocyte cultures exposed to γ-rays and bleomycin. Cancer Res 44:1499–1504

Morris DL, Ramanujam VMS, Legator MS (1986) Coumarin inhibits micronuclei formation in male ICR mice by benzo<a>pyrene and alters the pattern of urinary metabolites (Abstr) In: Shankel DM, Hartman PE, Kada T, Hollaender A (eds) Antimutagenesis and anticarcinogenesis mechanisms I. Plenum, New York, p 570

Moutschen J (1958) Cystamine as a modifier of chromosome breaks caused by maleic hydrazide in broad bean. Radiat Res 9:188

Moutschen J (1960) Modifications with cystamine of chromosome damage caused by maleic hydrazide in broad bean. Radiobiol Latina III:271–277

Müller D (1980) Validity of test systems used in the detection of mutagenic and carcinogenic properties of chemical substances. In: Norpoth KH, Garner RC (eds) Short-term test systems for detecting carcinogens. Springer, Berlin Heidelberg New York, pp 141–159

Murota T, Shibuya T (1983) Modifying effects of the enzyme inducers, phenobarbital and 3-methylcholanthrene, on dominant lethal events induced by 7,12-dimethylbenz<a>anthracene in mice. Mutation Res 107:329–336

Nagasawa H, Little JB (1981) Factors influencing the induction of sister chromatid exchanges in mammalian cells by 12-o-tetradecanoyl-phorbol-13-acetate. Carcinogenesis 2:601–607

Nair ND, Vogel R (1989) Chromosome aberration test and sister chromatid exchange assay in murine bone marrow cells after in vivo exposure to d-penicillamine alone or combined with cyclophosphamide. Mutation Res 227:237–240

Namiki M, Osawa T (1986) Antioxidants/antimutagens in food. In: Shankel DM, Hartman PE, Kada T, Hollaender A (eds) Antimutagenesis and anticarcinogenesis mechanisms I. Plenum Press, New York, pp 131–142

Nasjleti CE, Spencer HH (1969) Chromosome damage after treatment with cyteamine in non-irradiated and irradiated human leukocytes. J Nucl Med 10:495–500

Natarajan AT (1984) Origin and significance of chromosomal alterations. In: Obe G (ed) Mutations in man. Springer, Berlin Heidelberg, New York, pp 156–176

Natarajan AT, Obe G (1984) Molecular mechanisms involved in the production of chromosomal aberrations. III. Restriction endonucleases. Chromosoma (Berlin) 90:120–127

Natarajan AT, Simons JWIM, Vogel EW, Van Zealand AA (1984) Relationship between killing, chromosomal aberrations, sister chromatid exchanges and point mutations induced by monofunctional alkylating agents in Chinese hamster cells. A correlation with different ethylation products in DNA. Mutation Res 128:31–40

Natarajan AT, Rotteveel AHM, Van Pieterson J, Schliermann MG (1986) Influence of incorporated 5-bromodeoxyuridine on the frequencies of spontaneous and induced sister-chromatid exchanges, detected by immunological methods. Mutation Res 163:51–55

Neis J (1986) Variation in genotoxicity due to differences in metabolic activation. Thesis, Nijmegen, 170 pp

Nemirovsky LE, Klimenko VV (1973) Effect of caffeine on genetic lesions induced by dipine (an alkylating compound) in hepatocytes of rats. Genetika 9/6:100–106 (in Russian)

Nordenson I (1977) Effect of superoxide dismutase and catalase on spontaneously occurring chromosome breaks in patients with Fanconi's anemia. Hereditas 86:147–150

Norppa H, Westermarck T, Laasonen M, Knuutila L, Knuutila S (1980a) Chromosomal effects of sodium selenite in vivo. I. Aberrations and sister chromatid exchanges in human lymphocytes. Hereditas 93:93–96

Norppa H, Westermarck T, Oksanen A, Rimaila-Parnanen E, Knuutila S (1980b) Chromosomal effects of sodium selenite in vivo. II. Aberrations in mouse bone marrow and primary spermatocytes. Hereditas 93:97–99

Norppa H, Westermarck T, Knuutila S (1980c) Chromosomal effects of sodium selenite in vivo. III. Aberrations and sister chromatid exchanges in Chinese hamster bone marrow. Hereditas 93:101:105

Novick A, Szilard L (1952) Anti-mutagens. Nature (Lond) 170:926–927

Nowak C, Obe G (1984) On the origin of chromosomal aberrations in human peripheral lymphocytes in vitro. I. Experiments with Neurospora endonuclease and polyethylene glycol. Hum Genet 66:335–343

Nuzhdin NJ, Niznik GV (1963) The effects of protection against genetic damage caused by ionizing radiation in mammalian sex cells. Genetics Today 1:5.75

Obe G (1969) Die interzelluläre Verteilung chemisch induzierter Chromatidbrüche und -translokationen bei menschlichen Lymphocyten in vitro. Z Naturforsch 24b:1207–1208

Obe G, Beek B (1982) The human leukocyte test system. In: De Serres FJ, Hollaender A (eds) Chemical mutagens, vol 7. Plenum, New York, pp 337–400

Obe G, Beek B (1984) Human peripheral lymphocytes in mutation research. In: Obe G (ed) Mutations in man. Springer, Berlin Heidelberg New York, pp 177–197

Obe G, Slacik-Erben R (1973) Suppressive activity by fluoride on the induction of chromosome aberrations in human cells with alkylating agents in vitro. Mutation Res 19:369–371

Obe G, Sperling K, Belitz HJ (1972) Vergleichende Untersuchungen zur Wirkung von Äthyleniminen auf menschliche Leukozytenchromosomen in vitro. Arzneimittel-Forsch (Drug Res) 22:982–984

Obe G, Vogt HJ, Madle S, Fahning A, Heller WD (1982a) Double-blind study on the effect of cigarette smoking on the chromosomes of human peripheral blood lymphocytes in vivo. Mutation Res 92:309–319

Obe G, Kalweit S, Novak C, Ali-Osman F (1982b) Liquid holding experiments with human peripheral lymphocytes. I. Effects of liquid holding on sister chromatid exchanges induced by trenimon, diepoxybutane, bleomycin and X-rays. Biol Zbl 101:97–113

Obe G, Vasudev V, Johannes C (1987) Chromosome aberrations induced by restriction endonucleases. In: Obe G, Basler A (eds) Cytogenetics: basic and applied aspects. Springer, Berlin Heidelberg New York, pp 300–314

Ockey CH (1981) Methyl methane-sulphonate (MMS) induced SCEs are reduced by the BrdU used to visualize them. Chromosoma 84:243–256

Padma PR, Amonkar AJ, Bhide SV (1989) Antimutagenic effects of betel leaf extract against the mutagenicity of two tobacco-specific N-nitrosamines. Mutagenesis 4:154–156

Pantelias GE, Maillie HD (1984) The use of peripheral blood mononuclear cell prematurely condensed chromosomes for biological dosimetry. Radiat Res 99:140–150

Paschin YV, Bakhitova LM, Benthen TI (1986) Increased antimutagenic activity of simple substituted phenols mixed with the hindered phenolic antioxidant dibunol. Food Chem Toxicol 24:881–883

Perry P, Evans HJ (1975) Cytological detection of mutagen-carcinogen exposure by sister chromatid exchange. Nature (Lond) 258:121–125

Perry P, Wolff S (1974) New Giemsa method for differential staining of sister chromatids. Nature (Lond) 261:156–158

Phillips TL, Wasserman TH (1984) Promise of radiosensitizers and radioprotectors in the treatment of human cancer. Cancer Treat Rep 68:291–302

Pincheira J, Bravo M, Lopez-Saez JF (1988) Fanconi's anemia lymphocytes: effect of caffeine, adenosine and niacinamide during G2 prophase. Mutation Res 199:159–165

Pirozzi SJ, Renz JF, Kalf GF (1989) The prevention of benzene-induced genotoxicity in mice by indomethacin. Mutation Res 222:291–298

Pohl H, Reidy JA (1989) Vitamin C intake influences the bleomycin-induced chromosome damage assay: implications for detection of cancer susceptibility and chromosome breakage syndromes. Mutation Res 224:247–252

Popescu NC, Amsbaugh SA, Di Paolo JA (1980) Reduced N-methyl-N'-nitro-N-nitrosoguanidine sister chromatid exchange induction in Chinese hamster V79 cells pre-exposed to 5-bromodeoxyuridine. Chromosoma 76:329–338

Porfirio B, Ambroso G, Gianella G, Isacchi G, Dallapiccola B (1989) Partial correction of chromosomal instability in Fanconi anemia by desferrioxamine. Hum Genet 83:49–51

Preston RJ, Au W, Bender MA, Brewen JG, Carrano AV, Heddle JA, McFee AF, Wolff S, Wassom JS (1981) Mammalian in vivo and in vitro cytogenetic assays: a report of the U.S. EPA's Gene-Tox Program. Mutation Res 87:143–188

Purchase IFH (1978) Chromosomal analysis of exposed populations: a review of industrial problems. In: Evans HJ, Lloyd DC (eds) Mutagen-induced chromosome damage in man. Edinburgh Univ Press, Edinburgh, pp 258–267

Quin S, Batt T, Huang CC (1985) Influence of retinol on carcinogen-induced sister chromatid exchanges and chromosome aberrations in V79 Cells. Environ Mutagen 7:137–146

Raj AS, Heddle JA (1980) The effect of superoxide dismutase, catalase and l-cysteine on spontaneous and on mitomycin C induced chromosomal breakage in Fanconi's anemia and normal fibroblasts as measured by the micronucleus method. Mutation Res 78:59–66

Raj AS, Katz M (1983) Inhibitory effect of 7,8-benzoflavone on DMBA- and BaP-induced bone marrow micronuclei in mice. Mutation Res 110:337–342

Raj AS, Katz M (1984a) Inhibitory effects of α- and β-naphthoflavones on DMBA-induced anomalies in germ cells assessed by the sperm abnormality assay. Mutation Res 136:81–84

Raj AS, Katz M (1984b) Corn oil and its minor constituents as inhibitors of DMBA-induced chromosomal breaks in vivo. Mutation Res 136:247–253

Raj AS, Katz M (1985a) Glutathione as an inhibitor of chromosomal breakage induced by DMBA, BaP, CP, and MMC assessed by the in vivo bone marrow micronucleus test (Abstr). Environ Mutagen 7 (Suppl 3):52

Raj AS, Katz M (1985b) β-Carotene as an inhibitor of benzo<a>pyrene and mitomycin C induced chromosomal breaks in the bone marrow of mice. Can J Genet Cytol 27:598–602

Raj AS, Katz M (1986) Selenium as an inhibitor of mutagen-induced clastogenicity in mice tested by micronucleus assay (Abstr). Environ Mutagen 8 (Suppl 6):66–67

Ramel C, Alekperov UK, Ames BN, Kada T, Wattenberg LW (1986) Inhibitors of mutagenesis and their relevance to carcinogenesis. ICPEMC Publ NO 12. Mutation Res 168:47–65

Rao AP, Rao PN (1976) Premature chromosome condensation and cell cycle analysis. J Natl Cancer Inst 57:1139–1143

Rao KP, Ramadevi G, Das UN (1986) Vitamin A can prevent genetic damage induced by benzo<a>pyrene to the bone marrow cells of mice. Int J Tissue React 8:219–223

Raposa T, Sreter L, Pollak Z, Varkonyi J, Feher J (1982) SCE in human "metabolizing system". The assessment of DNA damage caused by cytostatic therapy and its modification by a dihydroquinoline-type antioxidant (Abstr). Mutation Res 97:216

Rau D, Gebhart E (1987) Studies on the activity of the radio-protective agent WR 2721 on the clastogenicity of trenimon in 3 cell systems (Abstr). Mutation Res 182:289

Ray JH, Altenburg LC, Jacobs MM (1978) Effect of sodium selenite and methylmethanesulfonate or N-hydroxy-2-acetylaminofluorene coexposure on sister chromatid exchange production in human whole blood cultures. Mutation Res 57:359–368

Renner HW (1984) Antimutagenic effect of an antioxidant in mammals. Mutation Res 135:125–129

Renner HW (1985) Anticlastogenic effect of β-carotene in Chinese hamsters. Time and dose response studies with different mutagens. Mutation Res 144:251–256

Renner HW (1986) The anticlastogenic potential of fatty acid methyl esters. Mutation Res 172:265–269

Renner HW (1990a) In vivo effects of single or combined dietary antimutagens on mutagen-induced chromosomal aberrations. Mutation Res 244:185–188

Renner HW (1990b) Anticlastogenic dietary factors assessed in mammalian assays. In: Kuroda Y, Shankes DM, Waters MD (eds) Antimutagenesis and anticarcinogenesis mechanisms II. Plenum, New York, pp 35–43

Renner HW, Knoll M (1984) Antimutagenic effects on male germ cells of mice. Mutation Res 140:127–129

Revazova YA, Radchenko LU (1976) The modifying action of polyvinylpyrrolidone on the mutagenic activity of photrine. Khim Farm Zh 10/9:11–15 (in Russian)

Revell SH (1955) A new hypothesis for "chromatid" changes. In: Bacq ZM, Alexander P (eds) Radiobiology Symp Liege 1954. Butterworths, London, pp 243–253

Revesz L (1969) Interplay between natural and artificial radiation protection. In: IAEA: radiation damage and sulphydryl compounds. Vienna, pp 125–132

Rieger R, Michaelis A (1967) Die Chromosomenmutationen. Fischer, Jena

Rieger R, Michaelis A, Nicoloff H (1982) Inducible repair processes in plant root tip meristems? "Below-additivity effects" of unequally fractionated clastogen concentrations. Biol Zbl 101: 125–138

Rieger R, Michaelis A, Nicoloff H (1986) Effects of stress factors on the clastogen response of *Vicia faba* root-tip meristems -- "clastogenic adaptation". Biol Zbl 105:19–28

Riley HP (1952) Preliminary report on the effect of certain chemicals on radiation damage to chromosomes. Genetics 37:618–619

Röhrborn G (1968) Chemische Konstitution und mutagene Wirkung. IV. Zyklische N-Lostderivate. Molec Gen Genet 102:50–68

Roots R, Okada S (1972) Protection of DNA molecules of cultured mammalian cells from radiation-induced single-strand scission by various alcohols and SH compounds. Int J Radiat Biol 21:329–342

Rosin MP, German J (1985) Evidence for chromosome instability in vivo in Bloom syndrome increased numbers of micronuclei in exfoliated cells. Hum Genet 71:187–191

Rosin MP (1990) Antigenotoxic activity of carotenoids in carcinogen-exposed populations. In: Kada Y, Shankel DM, Waters MD (eds) Antimutagenesis and anticarcinogenesis mechanisms II. Plenum, New York, pp 45–59

Rossman TG, Molina M (1983) Rat liver S9 preparations contain comutagenic activity. Mutation Res 122:103–109

Rössner P (1977) Mutagenic effect of sodium arsenite in Chinese hamster cell line Dede. Mutation Res 46:234–235

Rössner P, Cerná M, Pokorná D, Hajek V, Petr J (1988) Effect of ascorbic acid prophylaxis on the frequency of chromosome aberrations, urine mutagenicity and nucleolus test in workers occupationally exposed to cytostatic drugs. Mutation Res 208:149–153

Rutten AAJJL, Wilmer JWGM (1986) Effect of cigarette smoke condensate and norharman on the induction of SCEs by direct and indirect mutagens in CHO cells. Mutation Res 172:61–67

Saccone GTP, Pariza MW (1981) Enhancement of hepatic microsome-mediated bacterial mutagenesis by the rat liver soluble protein fractions. Mutation Res 88:135–145

Sandberg AA (1980, 1990) The chromosomes of human cancer and leukemia, 1st and 2nd edn. Elsevier, New York

Sandberg AA (1982) Sister chromatid exchange. Liss, New York

Sarto F, Finotto S, Giacomelli L, Mazzotti D, Tomanin R, Levis AG (1987) The micronucleus assay in exfoliated cells of the human buccal mucosa. Mutagenesis 2:11–17

Sasaki MS (1980) Chromosome aberration formation and sister chromatid exchange in relation to DNA repair in human cells. In: Generoso WM, Shelby MD, De Serres FJ (eds) DNA repair and mutagenesis in eukaryotes. Plenum, New York, pp 285–313

Sasaki MS (1982) Sister chromatid exchange as a reflection of cellular DNA repair. In: Sandberg AA (ed) Sister chromatid exchange. Liss, New York, pp 135–161

Sasaki YF, Imanishi H, Ohta T, Shirasu Y (1987) Effects of antimutagenic flavourings on SCEs induced by chemical mutagens in cultured Chinese hamster cells. Mutation Res 189:313–318

Sasaki YF, Imanishi H, Ohta T, Watanabe M, Matsumoto K, Shirasu Y (1988) Suppressing effect of tannic acid on UV and chemically induced chromosome aberrations in cultured mammalian cells. Agric Biol Chem 52:2423–2428

Sasaki YF, Imanishi H, Ohta T, Watanabe M, Matsumoto K, Shirasu Y (1989) Suppressing effect of tannic acid on the frequencies of mutagen-induced sister-chromatid exchanges in mammalian cells. Mutation Res 213:195–203

Sasaki YF, Imanishi H, Watanabe M, Ohta T, Shirasu Y (1990a) Suppressing effect of antimutagenic flavorings on chromosome aberrations induced by UV-light and X-rays in cultured Chinese hamster cells. Mutation Res 229:1–10

Sasaki YF, Matsumoto K, Imanishi H, Watanabe M, Ohta T, Shirasu Y, Tutikawa K (1990b) In vivo anticlastogenic and antimutagenic effects of tannic acid in mice. Mutation Res 244:43–47

Savage JRK (1976) Classification and relationships of induced chromosomal structural changes. J Med Genet 13:103–122

Savkovic N, Pecevski J (1973) A search for chromosome translocations induced by cysteamine and AET · Cl · HCl in spematogonia of irradiated mouse. Int J Radiat Biol 21:200–201

Sax K (1940) Analysis of X-ray induced chromosomal aberrations in *Tradescantia*. Genetics 25:41–68

Schmähl D (1976) Combination effects in chemical carcinogenesis (experimental results). Oncology 33:73–76

Schmid W (1976) The micronucleus test for cytogenetic analysis. In: Hollaender A (ed) Chemical mutagens. Principles and methods for their detection, vol 4. Plenum, New York, pp 31–53

Schneider EL, Lewis J (1982) Comparison of in vivo and in vitro SCE induction. Mutation Res 106:85–90

Schull WJ, Otake M, Neel JV (1981) Genetic effects of the atomic Bombs: a reappraisal. Science 213:1220–1227

Schulte PA, Ringen K, Hemstreet GP (1986) Optimal management of asymptomatic workers at high risk of bladder cancer. J Occup Med 28:13–17

Schwartz JL, Morgan WF, Brown-Lindquist P, Afzal V, Weichselbaum RR, Wolff S (1985) Comutagenic effects of 3-aminobenzamide in Chinese hamster ovary cells. Cancer Res 45:1556–1559

Schwartz JL, Weichselbaum RR (1985) Antipain-mediated suppression of sister chromatid exchanges induced by an inhibitor of poly (ADP-ribose) polymerase. Environ Mutagen 7:703–709

Schwarz K (1976) Essentiality and metabolic functions of selenium. Med Clin N Am 60:745–758

Seleznev YV (1976) Combined effect of N-nitroso-N-methylurea and actinomycin D on Chinese hamster chromosomes. Genetika 12/3:66–68 (in Russian)

Sen P, Hittelman WN (1984) Kinetics and extent of repair of bleomycin-induced chromosome damage in quiescent normal human fibroblasts and human mononuclear blood cells. Cancer Res 44:591–596

Sen NP, Smith DC, Schwinghammer L (1969) Formation of N-nitrosamines from secondary amines and nitrites in human and animal gastric juice. Food Cosmet Toxicol 7:301–307

Shamberger RJ (1984) Genetic toxicology of ascorbic acid. Mutation Res 133:135–139

Shamberger RJ, Baughman FF, Kalchert SL, Willis CE, Hoffman GC (1973) Carcinogen-induced chromosomal breakage decreased by antioxidants. Proc Natl Acad Sci USA 70:1461–1463

Shankel DM, Hartman PE, Kada T, Hollaender A (eds) (1986) Antimutagenesis and anticarcinogenesis mechanisms. Plenum, New York

Shannon RE (1975) Systems simulation: the art and science. Prentice-Hall, Englewood Cliffs, NJ

Shapiro B, Kollmann G (1969) Mechanisms of protection of macromolecules against ionizing radiation by sulphydryl and other protective agents. In: IAEA: radiation damage and sulphydryl compounds. Vienna, pp 23–43

Shcheglova EG, Chebotarev AN (1983) Comparison of level of sister chromatid exchanges and chromosomal aberrations induced by chemical mutagens in vitro. Bull Exper Biol Med 96:1604–1606

Simic MG (1988) Mechanisms of inhibition of free-radical processes in mutagenesis and carcinogenesis. Mutation Res 202:377–386

Sirianni SR, Chen HH, Huang CC (1981) Effect of retinoids on plating efficiency, sister-chromatid exchange (SCE) and mitomycin-C-induced SCE in cultured Chinese hamster cells. Mutation Res 90:175–182

Slacik-Erben R, Obe G (1976) The effect of sodium fluoride on DNA synthesis, mitotic indices and chromosomal aberrations in human leukocytes treated with trenimon in vitro. Mutation Res 37:253–266

Slaga TJ (1989) Cellular and molecular mechanisms involved in multistage skin carcinogenesis. In: Conti CJ, Slaga TJ, Klein-Szanto AJP (eds) Skin tumors: experimental and clinical aspects. Raven, New York, pp 1–18

Smalls E, Pettersen RM (1982) Reduction of benzo<a>pyrene induced chromosomal aberrations by dl-alpha-tocopherol. Eur J Cell Biol 26:96–97

Smetaná K, Lejnar J, Potmešil M (1969) A further contribution to the demonstration of RNA and nucleoli of blood cells in smear preparations. Folia Haematol 91:381–386

Smetaná K, Lejnar J, Salková J (1980) Studies on nucleoli in rosetting T and B lymphocytes of the human peripheral blood. Folia Haematol 107:720–723

Sobels FH (1980) Evaluating the mutagenic potential of chemicals: the minimal battery and extrapolation problems. Arch Toxicol 47:21–30

Sognier MA, Hittelman WN, Rao PN (1979) Effect of DNA repair inhibitors on the induction and repair of bleomycin-induced chromosome damage. Mutation Res 60:61–72

Sono A, Sakaguchi K (1981a) Inhibition of protein synthesis antagonizes induction of sister chromatid exchanges by exogenous agents. Mutation Res 80:121–131

Sono A, Sakaguchi K (1981b) The suppression by protein synthesis inhibitors of sister chromatid exchanges induced by exogenous agents in the euchromatin and heterochromatin of Chinese hamster cells. Cell Struct Funct 6:209–216

Sono A, Sakaguchi K (1982) Effects of colchicine, colcemid, and vincristine on ethyl-methanesulfonate-induced sister chromatid exchanges and chromosomal aberrations in Chinese hamster cells. Gann 73:912–919

Sorsa M (1984) Monitoring of sister chromatid exchange and micronuclei as biological endpoints. In: Berlin A, Draper M, Hemminki K, Vainio H (eds) Monitoring human exposure to carcinogenic and mutagenic agents. IARC Sci Publ No 59:339–349

Sorsa M, Vainio H (eds) (1982) Mutagens in our environment. Liss, New York

Sorsa M, Hemminki K, Vainio H (1982) Biologic monitoring of exposure to chemical mutagens in the occupational environment. Teratogen Carcinogen Mutagen 2:137:150

Speit G, Vogel W (1982) The effect of sulfhydryl compounds on sister chromatid exchanges. II. The question of cell specificity and the role of H_2O_2. Mutation Res 93:175–183

Speit G, Wick C, Wolf M (1980a) Induction of sister chromatid exchanges by hydroxylamine, hydrazine and isoniazid, and their inhibition by cysteine. Hum Genet 54:155–158

Speit G, Wolf M, Vogel W (1980b) The SCE inducing capacity of vitamin C: investigations in vitro and in vivo. Mutation Res 78:273–278

Speit G, Mehnert K, Wolf M, Vogel W (1982) Ultraviolet-induced sister chromatid exchanges in V79 cells with normal and BrdUrd substituted DNA and the influence of intercalating substances and cysteine. Radiat Res 90:538–546

Speit G, Hochsattel R, Vogel W (1984) The contribution of DNA single-strand breaks to the formation of chromosome aberrations and SCEs. In: Tice RR, Hollaender A (eds) Sister chromatid exchanges. Plenum, New York, pp 229–243

Sperling K (1984) Frequency and origin of chromosome abnormalities in man. In: Obe G (ed) Mutations in man. Springer, Berlin Heidelberg New York, pp 128–146

Spitkovskii DM, Andrianov VT, Pisarevskii AN (1969) Radiation biophysics of nucleoproteid. Atomizdat, Moscow (in Russian)

Srám RJ, Kocisová J (1975) Effect of antibiotics on the mutagenic activity induced by chemicals. I. Chromosome aberrations during spermatogenesis in mice. Folia Biol 21:50–59

Šrám RJ, Kuleshov NP (1980) Monitoring the occupational exposure to mutagens by the cytogenetic analysis of human peripheral lymphocytes in vivo. Arch Toxicol Suppl 4:11–18

Šrám RJ, Samková I, Holá N (1983a) High-dose ascorbic acid prophylaxis in workers occupationally exposed to halogenated ethers. J Hyg Epid Microbiol Immunol 27:305–318

Šrám RJ, Dobiáš L, Pastorková A, Rössner P, Janca L (1983b) Effect of ascorbic acid prophylaxis on the frequency of chromosome aberrations in the peripheral lymphocytes of coal-tar workers. Mutation Res 120:181–186

Šrám RJ, Holá N, Kotešovec F, Vávra R (1985) Chromosomal abnormalities in soft coal open-cast mining workers. Mutation Res 144:271–275

Šrám RJ, Cerná M, Holá N (1986) Effect of ascorbic acid prophylaxis in groups occupationally exposed to mutagens. In: Ramel C, Lambert B, Magnusson J (eds) Genetic toxicology of environmental chemicals. Part B: Genetic effects and applied mutagenesis. Liss, New York, pp 327–335

Stadler LJ (1932) On the genetic nature of induced mutations in plants. In: Proc 6th Int Congr Genet 1:274–294

Steffen JA, Swierkowska K, Michalowski A, Kling E, Nowakowska A (1978) In vitro kinetics of human lymphocytes activated by mitogens. In: Evans HJ, Lloyd DC (eds) Mutagen-induced chromosome damage in man. Univ Press, Edinburgh, pp 89–107

Steinheider G, Neth R, Marquardt H (1985) Evaluation of nongenotoxic and genotoxic factors modulating the frequency of micronucleated erythrocytes in the peripheral blood of mice. Cell Biol Toxicol 1:197–211

Stevenson AC, San Roman C, Patel CR (1973) Effects of amylobarbitone on the frequency of chromosomal aberrations in human lymphocytes determined by chlorambucil and melphalan in vitro. Mutation Res 19:225–119

Stich HF (1986) Reducing the genotoxic damage in the oral mucosa of betel quid/tobacco chewers. In: Shankel DM, Hartman PE, Kada T, Hollaender A (eds) Antimutagenesis and anticarcinogenesis mechanisms. Plenum, New York, pp 381–391

Stich HF, Dunn BP (1986) Relationship between cellular levels of beta-carotene and sensitivity to genotoxic agents. Int J Cancer 38:713–717

Stich HF, Rosin MP (1985) Towards a more comprehensive evaluation of a genotoxic hazard in man. Mutation Res 150:43–50

Stich HF, Curtis JR, Parida BB (1982a) Application of the micronucleus test to exfoliated cells of high cancer risk groups: tobacco chewers. Int J Cancer 30:553–559

Stich HF, Stich W, Parida BB (1982b) Elevated frequency of micronucleated cells in the buccal mucosa of individuals at high risk for oral cancer: betel quid chewers. Cancer Lett 17:125–134

Stich HF, Stich W, Rosin MP, Vallejera MO (1984) Use of the micronucleus test to monitor the effect of vitamin A, beta-carotene and canthaxanthin on the buccal mucosa of betel nut/tobacco chewers. Int J Cancer 34:745–750

Stich HF, Hornby AP, Dunn BP (1985) A pilot beta-carotene intervention trial on Inuits using smokeless tobacco. Int J Cancer 36:312–327

Stosiek M, Gebhart E (1974) Protective effect of Reducdyn^{R} against the chromosome damaging activity of 2,3,5-triethyleneiminebenzoquinone-1,4 on human lymphocytes in vitro. Humangenetik 25:209–216

Stukalov SV, Chebotarev AN (1983) Comparative study of the rate of sister chromatid exchanges induced by thiophosphamide in vivo and in vitro. Bull Exp Biol Med 95:74–75

Sugimura T (1982) Mutagens, carcinogens, and tumor promoters in our daily food. Cancer 49:1970–1984

Sutherland GR (1983) The fragile X chromosome. Int Rev Cytol 81:107–143

Sutherland GR, Hecht F (1985) Fragile sites on human chromosomes. Oxford Univ Press, New York

Sutherland GR, Simmers RN (1988) No statistical association between common fragile sites and nonrandom chromosome breakpoints in cancer cells. Cancer Genet Cytogenet 31:9–15

Suzuki N (1986) Effects of human interferon-α on UV-induced DNA-repair synthesis and cell killing. Mutation Res 175:189–196

Suzuki S, Atai H, Hatakeyama Y, Hara M, Nakagawa S (1989) Administration-route-related differences in the micronucleus test with benzene. Mutation Res 223:407–410

Sweins A (1983) Protective effect of selenium against arsenic-induced chromosomal damage in cultured human lymphocytes. Hereditas 98:249–252

Takeda Y, Sugahara T (1960) The effects of AET (s,2-aminoethylisothiourea · Br · HBr) on the induction of dominant lethal mutations by X-rays. Nipp Acta Radiol 20:1996

Tates AD, Neuteboom I, Hofker M, Den Engelse L (1980) A micronucleus technique for detecting clastogenic effects of mutagens/carcinogens (DEN, DMN) in hepatocytes of rat liver in vivo. Mutation Res 74:11–20

Tates AD, Dietrich AJJ, de Vogel N, Neuteboom I, Bos A (1983) A micronucleus method for detection of meiotic micronuclei in male germ cells of mammals. Mutation Res 121:131–138

Thompson JS, Thompson MW (1986) Genetics in medicine. 4th ed. Saunders, Philadelphia

Tice RR, Hollaender A (eds) (1984) Sister chromatid exchanges. A and B. Plenum, New York

Tittelbach H, Brüderlein S, Gebhart E (1989) The activity of bleomycin and modifiers of its clastogenic effect on the interphase chromatin of Indian muntjak fibroblasts. Cell Biol Toxicol 5:39–50

Tobey RA (1985) Different drugs arrest cells at a number of distinct stages in G2. Nature (Lond) 254:245–247

Tohda H, Oikawa A (1986) Proliferation-dependent reduction of sister-chromatid exchange frequency induced by mitomycin C in human lymphoblastoid cells and its suppression by inhibitors of DNA replication. Mutation Res 163:167–174

Tohda H, Sugawara R, Tazawa J, Tada M, Oikawa A (1984) Antagonizing effect of 3-aminoharman on induction of sister-chromatid exchanges by mutagens. Mutation Res 129:63–69

Tosch R (1960) Über klinische Erfahrungen mit organischen Sulfhydrylverbindungen in der Strahlentherapie. Fortschr Med 78:511

Trush MA, Mimnaugh EG, Ginsburg E, Gram TE (1982) Studies on the interaction of bleomycin A2 with rat lung microsomes. J Pharmacol Exp Ther 221:152–159

Trzos RJ, Petzold GL, Brunden MN, Swenberg JA (1978) The evaluation of sixteen carcinogens in the rat using the micronucleus test. Mutation Res 58:79–86

Tsuchimoto T, Matter BE (1979) In vivo cytogenetic screening methods for mutagens with special reference to the micronucleus test. Arch Toxicol 42:239–248

Turrisi AT, Glover DJ, Hurwitz S, Glick J, Norfleet AL, Weiler C, Yuhas JM, Kligerman MM (1986) Final report of the phase I trial of single-dose WR-2721 <S-2-(3-aminopropylamino) ethylphosporothioic acid>. Cancer Treat Rep 70:1389–1393

Umezawa K, Sawamura M, Matsushima T, Sugimura T (1979) Inhibition of chemically induced sister chromatid exchanges by elastatinal. Chem Biol Interact 24:107–110

Umezawa K, Sawamura M, Matsushima T, Sugimura T (1980) Inhibition of chemically induced sister chromatid exchanges by protease inhibitors from *Streptomyces*. Chem Biol Interact 30:247–251

Upton AC (1986) Evolving perspectives on the biology and mechanisms of carcinogenesis. Leukemia Res 10:727–734

Van Went GF (1979) Investigation into the mutagenic activity of azathioprine (Imuran^{R}) in different test systems. Mutation Res 68:153–162

Veltischev YE, Seleznev YV (1978) Antimutagenic effect of a hydroxyethylidendiphosphonic acid derivative. Genetika 14/7:1276–1277

Viktorov VV (1985) The imitation modeling of the action of kinetics of cell proliferation on the frequency and distribution of chromosomal aberrations in cells. Inst Med Genet Acad Med Sci USSR, Moscow, p 22 (in Russian)

Vogel E, Natarajan AT (1982) The relation between reaction kinetics and mutagenic action of monofunctional alkylating agents in higher eukaryotic systems: interspecies comparisons. In: De Serres FJ, Hollaender A (eds) Chemical mutagens. Principles and methods for their detection, vol 7. Plenum, New York, pp 295–336

Vogel F, Vrba M (1967) Influence of chloramphenicol on the reunion frequency of cytoxan-induced chromosome breaks in HeLa cells. Mutation Res 4:874–875

Vogel F, Röhrborn G, Hansmann I (1974) Die Testung von Fremdstoffen auf Mutagenität. Arzneimittel-Forsch (Drug Res) 24:1665–1677

Vogel F, Motulsky AG (1986) Human Genetics. 2nd ed. Springer, Berlin, Heidelberg, New York etc.

Vos O, Kalen MCAC (1968) Protection against ionizing radiation at the cellular level, assessed by various parameters. Int J Radiat Biol 14:107–118

Wang ZY, Cheng SJ, Zhou ZC, Athar M, Khan WA, Bickers DR, Mukhtar H (1989) Antimutagenic activity of green tea polyphenols. Mutation Res 223:273–285

Wargovich MG (1987) Diallyl sulfide, a flavor component of garlic (*Allium sativum*) inhibits dimethylhydrazine induced colon cancer. Carcinogenesis 8:487–489

Watanabe T, Endo A, Kato Y, Shima S, Watanabe T, Ikeda M (1980) Cytogenetics and cytokinetics of cultured lymphocytes from benzene-exposed workers. Int Arch Occup Environ Health 46:31–41

Waters MD, Brady AL, Stack HF, Brockman HE (1990) Antimutagenicity profiles for some model compounds. Mutation Res 238:57–85

Wattenberg LW (1983) Inhibition of neoplasia by minor dietary constituents. Cancer Res 43 (Suppl):2448–2459

Wei L, Whiting RF, Stich HF (1981) Inhibition of chemical mutagenesis: an application of chromosome aberration and DNA synthesis assays using cultured mammalian cells. In: Stich HF, San RHC (eds) Short-term tests for chemical carcinogens. Springer, Berlin Heidelberg New York, pp 428–437

Weinstein IB, Gattoni-Celli S, Kirschmeier P, Lambert M, Hsiao W, Backer J, Jeffrey A (1984) Multistage carcinogenesis involves multiple genes and multiple mechanisms. J Cell Physiol Suppl 3:127–137

Whiting RF, Wei L, Stich HF (1981a) Chromosome-damaging activity of ferritin and its relation to chelation and reduction of iron. Cancer Res 41:1628–1636

Whiting RF, Wei L, Stich HF (1981b) Mutagenic and antimutagenic activities of selenium compounds in mammalian cells. In: Spallholz JL, Martin JL, Ganther HE (eds) Selenium in biology and medicine. AVI, Westport, CN, pp 325–330

Whorton EB, Tice RR, Stetka DG (1984) Statistical design, analysis and inference issues in studies using sister chromatid exchange. In: Tice RR, Hollaender A (eds) Sister chromatid exchanges. Pt A. Plenum, New York, pp 431–440

Wiencke JK, Kelsey K, Kreiger R, Garry VF (1984) Serum- and plasma-dependent variations of benzo(α)pyrene-induced sister chromatid exchange in human lymphocytes. In: Tice RR, Hollaender A (eds) Sister chromatid exchanges. Pt B, Plenum, New York, pp 715–728

Wild D (1978) Cytogenetic effects in the mouse of 17 chemical mutagens and carcinogens evaluated by the micronucleus test. Mutation Res 56:319–327

Wilmer JL, Erexson GL, Kligerman AD (1986) Attenuation of cytogenetic damage by 2-mercaptoethanesulfonate in cultured human lymphocytes exposed to cyclophosphamide and its reactive metabolites. Cancer Res 46:203–210

Wolff S (1978) Chromosomal effects of mutagenic carcinogens and the nature of the lesions leading to sister chromatid exchange. In: Evans HJ, Lloyd DC (eds) Mutagen-induced chromosome damage in man. Univ Press, Edinburgh, pp 208–215

Wolff S (1980) Repair and induction of chromosome aberrations and point mutations in mammalian somatic cells: a summary and perspective. In: Generoso WM, Shelby MD, De Serres FJ (eds) DNA repair and mutagenesis in eukaryotes. Plenum, New York, pp 367–371

Wolff S, Arutyunyan R (1979) The apparent decrease in thio-TEPA-induced chromosome aberrations in human lymphocytes caused by an effect of WR 2721 on the cell cycle as found by the definitively determined division method. Environ Mutagen 1:5–13

Wolff S, Bodycote J, Painter RB (1974) Sister chromatid exchanges induced in Chinese hamster cells by UV irradiation at different stages of the cell cycle: the necessity of cells to pass through S. Mutation Res 25:73–81

Wolff S, Afzal V, Lindquist PB (1984) Cultured human lymphocytes proliferate faster in medium lacking fetal calf serum and antibiotics. Mutation Res 129:207–213

Wolff S, Olivieri G, Afzal V (1990a) Differential adaptive response of human lymphocytes to different chemical mutagens (Abstr). Environ Molec Mutagen 15 (Suppl 17):45

Wolff S, Olivieri G, Afzal V (1990b) Adaptation of human lymphocytes to radiation or chemical mutagens: differences in cytogenetic repair. In: Obe G, Natarajan AT (eds) Chromosomal aberrations: basic and applied aspects. Springer, Berlin Heidelberg New York, pp 140–150

Yakovenko KN, Platonova VI (1979) Spontaneous level of sister chromatid exchanges and their distribution in human cells. Genetika 15:1115–1123 (in Russian)

Yakovenko KN, Sapacheva VA (1984) The statistical analysis of elimination of chromosome type aberrations and the fate of aberrant cells. Genetika 20:144–154 (in Russian)

Yakovenko KN, Azhayev SA, Bochkov NP (1974) Cytogenetic effect of ethyleneimine in a culture of human lymphocytes. II. A mathematical model of the effect of different concentrations of dipine and photrin. Genetika 10:138–146 (in Russian)

Yakovenko KN, Bochkov NP, Chebotarev AN (1976) The distribution of chemically induced chromosome breaks in human cells. Biol Zbl 95:437–450

Yakubovskaya EL, Makedonov GP, Zasukhina GD (1988) Natural and recombinant interferons protect the human cells with disorder in the repair system (Marfan syndrome) against the action of mutagens. Genetika 24:2050–2055 (in Russian)

Yunis JJ (1983) The chromosomal basis of human neoplasia. Science 221:227–236

Yunis JJ, Soreng AL, Bowe AE (1987) Fragile sites are targets of diverse mutagens and carcinogens. Oncogene 1:59–69

Zasukhina GD, Barkova EP, Andzhaparidze OG (1979) Effect of interferon in vitro on development of chromosomal aberrations and mitotic activity of human diploid cells after the action of 4-nitroquinoline-l-oxide. Vopr Virusol 5:544–547

Zasukhina GD, Sinelschikova TA, Vasilyeva IM, Kirkova ZS (1981) An approach to modification of the repair activity in eucaryotic cells by biological factors. Mutation Res 81:123–132

Zenzes MT (1987) Chromosome analyses using human spermatozoa. In: Obe G, Basler A (eds) Cytogenetics. Springer, Berlin Heidelberg New York, pp 198–224

Zimmering PH, Legator MS, Weber E (1975) The effect of pyrogen treatment on cytogenetic response with chemical mutagens in mice (Abstr). Mutation Res 31:339–340

Zollino M, Neri G (1987) Protective effect of S-adenosylmethionine against the induction of chromosome fragile sites. Acta Haematol 78 (Suppl 1):100–103

Subject Index

Printing: COLOR-DRUCK DORFI GmbH, Berlin
Binding: Buchbinderei Lüderitz & Bauer, Berlin